MW01493222

# Boxwood Handbook

# Boxwood Handbook

## A Practical Guide to Knowing and Growing Boxwood

by Lynn R. Batdorf
U.S. National Arboretum

The American Boxwood Society

## Dedication

To Ethel and Kenneth Peters, who exerted a major influence in guiding my early years in ornamental horticulture.

*L. R. B.*

The American Boxwood Society
P.O. Box 85, Boyce, VA 22620. All rights reserved.
First edition published 1995; revised 1997
Printed in the United States of America
Messenger Printing Company, St. Louis, Mo.

**Library of Congress Cataloging in Publication Data**
Batdorf, Lynn R.
      Boxwood Handbook/Lynn R. Batdorf. - 1st ed. rev. p. cm.
      Includes bibliographical references.
      ISBN 1-886833-00-1
      1. Horticulture
   Dewey decimal    1994    SB435
   CIP 95-75228

Photographs: Lynn R. Batdorf
Typography: John S. McCarthy
Hardiness Map: Agricultural Research Service, USDA
Front Cover: *Buxus sempervirens* 'Henry Shaw'
Back Cover: *Buxus sinica* var. *insularis* 'Wintergreen' at the Missouri Botanical Garden, St. Louis, Mo.

*This book is printed with soy ink on acid-free, recycled paper.*

# Contents

# Foreword

This publication is designed to serve as a practical guide to identify, propagate, care for and enjoy boxwoods that grow in the temperate zones of the world. It is the result of a cooperative effort between the U.S. Department of Agriculture (USDA), Agricultural Research Service (ARS), U.S. National Arboretum (USNA) and The American Boxwood Society (ABS).

Trade names are necessary to report factually on available data; however, the USDA and ABS neither guarantee nor warrant the standard of the product, and the use of trade names implies no approval of the products to the exclusion of others that may also be suitable.

# Acknowledgements

The author gratefully acknowledges the critical reading and editing of the manuscript as well as numerous suggestions for improvement, by the following individuals:

Scott Aker, USNA, Washington, D.C.
Albert Beecher, deceased, Virginia Polytechnic Institute and State University, Blacksburg, Va.
Raymond Bosmans, Cooperative Extension Service, Ellicott City, Md.
Joan & Scot Butler, ABS, Winchester, Va.
John Davidson, Retired, Department of Entomology, University of Maryland, College Park, Md.
Kristen Dill, USNA, Washington, D.C.
Ethel Dutky, Department of Plant Pathology, University of Maryland, College Park, Md.
Decca Frackelton, ABS, Fredericksburg, Va.
William Graves, Department of Horticulture, Iowa State University, Ames, Iowa
Mary Holekamp, ABS, Port Huron, Mich.
Richard Mahone, Retired, Colonial Williamsburg Foundation, Williamsburg, Va.
Peter Mazzeo, Retired, USNA, Washington, D.C.
Sylvester March, Retired, USNA, Washington, D.C.
Dean Norton, Mount Vernon Ladies Association, Mt. Vernon, Va.
Stephen Southall, ABS, Lynchburg College, Lynchburg, Va..

# Chapter 1
# History of Boxwood

Buxus sempervirens *'Suffruticosa' in the Upper Garden at Mount Vernon. Washington's greenhouse is the brick building on the left.*

# HISTORY

The American Boxwood Society proclaims boxwood as man's oldest garden ornamental. Boxwood has enriched our lives for centuries as a landscape plant, a source of lumber, and through its medicinal qualities. Leaf and fruit fossils from boxwood have been located in more than 20 separate locations throughout Europe.

The first known use of boxwood was ornamental and occurred among the Egyptians about 4,000 B.C. Boxwood was planted in their gardens and kept clipped into formal hedges. In 1,000 B.C. Homer wrote in the *Iliad* of boxwood used as yokes for the stallions driven by the King of Troy. There are references to boxwood in the Old Testament of the Bible. An 8th-century B.C. table made of boxwood with juniper inlay and a walnut top was recovered nearly intact during the discovery of the tomb of King Midas in 1957.

The first scholarly reference to boxwood was by Theophrastus, who lived in Greece from 372 to 287 B.C. He was a horticulturist who recognized the virtue of "box" as a wood product. During the reign of Emperor Augustus in the first century B.C., the villas of many affluent Romans were landscaped with formal boxwood plantings, usually in topiary form. Of all the early Roman gardeners, the best known was Pliny. He not only had boxwood in his garden but used it as a material for musical instruments as well.

The history of boxwood in England has been confusing in the past. In fact, workers have recovered pollen grains in England that are estimated to be 7,000 years old and charcoal remains dating to 2,000 B.C. The pollen grains show that boxwood grew in England prior to the last glacial period. All of the original boxwood were removed by this glacial epoch. It was the Romans during the Roman Empire who can be credited with reestablishing boxwood to England. Since then, boxwood has become naturalized in some areas. The best example of this is the now wild growth of boxwood shrubs on Box Hill in Dorking, Surrey. Boxwood has also been incorporated into many English gardens.

The first planting in North America is believed to have been by Nathaniel Sylvester who built a manor on a Long Island plantation in 1652. Soon after, plantings of *Buxus sempervirens* were made. Today, in North America, the uses of boxwood are mostly ornamental. Classic examples of boxwood can be seen in numerous faithful replicas of colonial gardens. Contemporary plantings of boxwood are still dominated by *B. sempervirens* 'Suffruticosa'. However, as the gardening public becomes better informed, the benefits and beauty of many other cultivars are increasingly appreciated.

# THE WOOD OF BOXWOOD

Boxwood has remarkable wood characteristics. The wood of *B. sempervirens* is quite heavy and dense. When newly cut it weighs 80 pounds 7 ounces per cubic foot and when fully dry is still a hefty 68 pounds 12 ounces. As a result, the timber of boxwood is quite valuable. The ancient Greeks and Romans found many uses for the wood: writing tablets, flutes, spinning tops. combs, jewel cases, carved ornaments and images, inlays and veneers.

When the power loom was invented, a demand developed for wood with great strength, elasticity, firmness, and uniformity, all qualities which boxwood has. From 1860 to 1880 the imports into England from the Caucasus, Asia Minor and Persia averaged about 6,000 tons of boxwood annually.

# BOXWOOD PLANTS

In 1753, when Carolus Linnaeus, the great Swedish botanist, published his famous book, *Species Plantarum*, boxwood was given the botanical name *Buxus sempervirens*. The Latin word *Buxus* means "box;" *sempervirens* means "evergreen."

Boxwood is a broad-leaved, evergreen shrub. Its habit can be dense or open and may have a single or multibranched trunk. the leaves are opposite, short-petioled, simple, with a leathery texture (coriaceous) and usually without hairs (glabrous). The yellow flowers are quite small and appear in early spring, usually in March. They have no petals and they are located in the axillary or terminal clusters,

*Boxwood flowers showing numerous yellow stamens surrounding the green central pistil.*

consisting usually of a single, terminal female flower that is surrounded by numerous male flowers.

The male (staminate) flowers have four sepals and four stamens that are much longer than the sepals. The female (pistillate) flowers have six sepals and a three-celled ovary with three short styles. The fruit (capsule) is ovoid in shape. It has three horns, and the capsule can be separated into three pieces, each of which has two horns. Each piece has two black shiny seeds, with a total of six seeds per capsule.

*The horns are clearly visible on these immature green capsules.*

# Chapter 2
# Recommended Boxwood Plants

Buxus sempervirens *and* B. sempervirens *'Suffruticosa' at Colonial Williamsburg in Virginia.*

# RECOMMENDED BOXWOOD PLANTS

The most common criticism of boxwood is its relative "sameness" exhibited by numerous cultivars. This arose from the fact that most homeowners grow only *Buxus sempervirens* 'Suffruticosa', the so-called "English" boxwood. In reality there are nearly 160 registered cultivars of boxwood.

Of this 160, there are about 115 different cultivars and species available commercially. A partial list of available boxwood cultivars follows and points out the unique characteristics that differentiate them from each other.

### *Buxus balearica* Lam.

*History:* Native of the mountainous regions of the three Balearic Islands off the east coast of Spain. It was discovered in 1785. Large quantities were exported from Istanbul, Turkey, for timber in the 1840s.

*Description:* A handsome species with impressive and massive, bright green, leathery leaves that are three times as large as *B. sempervirens*, averaging from 1¹/₂" to 2" long and ³/₄" to 1" wide.

*Size:* An 80-year old plant growing at the Royal Botanic Garden, Kew, London achieved a height of 30' with a single trunk 3'3" in girth. In southern North America, 8' to 10' tall and 3' to 4' wide is common.

*Zone:* Not as hardy as most temperate zone boxwoods, but does survive in Zones 8 to 10.

### *Buxus harlandii* Hance

*History:* A unique species discovered in China in October 1858.

*Description:* There are two clones of this species in existence. The first is hardy, with long and narrow leaves averaging 1" to 1³/₄" long and ³/₈" to ¹/₂" wide,

Buxus harlandii *Hance.*

*Recommended Plants*

with a low mounding habit. The second clone has leaves that are dark green, shiny and broad, averaging from $1^1/_2$" to $1^5/_8$" long and $^3/_8$" to $^5/_8$" wide. This tender plant has an interesting V shape, with a flat top the result of frost damage burning the early spring growth. Both clones have leaf tips which are retuse or indented.

*Size:* A 32-year old hardy plant is 33" tall and 6' wide. A 21-year-old tender plant is $4^1/_2$' tall and 3' wide. Both are at the U.S. National Arboretum.

*Zone:* The first clone is cold hardy to Zone 7. The second, more tender clone will grow in Zone 8.

### *Buxus harlandii* 'Richard'

Buxus harlandii *'Richard'*.

*History:* Originated as a mutation of *B. harlandii* at the Straughn Nursery, Louisiana in 1956.

*Description:* The dark green, heart-shaped leaf is quite shiny and is deeply emarginate. The plant has a vase shape with a flat top from frost pruning.

*Size:* Mature size is variable, averaging 3' tall and $2^1/_2$' wide

*Zone:* Hardy to Zone 8.

### *Buxus microphylla* Sieb. & Zucc.

*History:* The first boxwood from Asia to reach the Western world, it was introduced in 1860.

*Description:* An attractive small mound, the leaf is medium green with a yellow undertone which is characteristic of the species. Leaf shape is obovate to oblong-obovate, as frequently occurs in boxwoods of Asian origin.

This species, while not widely grown, has given rise to many superior, dwarf-growing cultivars.

*Size:* Mature height varies from 3' to 7'.

*Zone*: Hardy to Zone 5.

*Recommended Plants*

### *Buxus microphylla* 'Compacta'

*History:* Grown since 1912, it is occasionally misnamed 'Kingsville Dwarf' because it was first distributed by the Kingsville Nursery in Maryland and because it is a dwarf plant. Due to its dwarf nature it is often used as a bonsai plant. Sports of 'Compacta' have created four other plants: 'Curly Locks', 'Grace Hendrick Phillips', 'Helen Whiting', and 'Sunlight'. The sporting characteristic can become aggressive and should be pruned out to maintain the plant.

*Description:* Annual growth averages ¼" to ½" which makes this cultivar the slowest-growing of boxwoods. The leaves are quite small, averaging ½" long and less than ¼" wide. The plant has a tight, low mounding habit and grows best in full shade.

*Size:* The smallest of all boxwood cultivars, 25-year old plants average 10" in height and 18" in width.

*Zone:* Hardy to Zone 5.

### *Buxus microphylla* 'Curly Locks'

*History:* Originated as a sport from 'Compacta' in 1942.

*Description:* The small leaves have a twisting or curling habit. It is a medium-sized plant with a broad, spreading and open habit.

*Size:* At 30 years this plant will average 6' tall and 12' wide.

*Zone:* Hardy to Zone 5.

Buxus microphylla *'Curly Locks'*.

***Recommended Plants***

### Buxus microphylla 'Grace Hendrick Phillips'

*History:* Originated as a sport from 'Compacta' in 1959 at the Kingsville Nursery.

*Description:* A handsome, broadly conical, dwarf-growing plant with small dark green leaves throughout the year.

B. microphylla *'Grace Hendrick Phillips'*.

*Size:* At 21 years of age it will grow to 23" tall and 35" wide.

*Zone:* Hardy to Zone 5.

### Buxus microphylla 'Green Pillow'

*History:* Originated as an open-pollinated seedling in 1912.

*Description:* Similar to *B. microphylla* 'Compacta' except the leaves are about twice as large, the dense and compact habit makes it well suited as a border or an edging plant.

*Size:* At 30 years of age the plant is 30" high and 40" wide.

*Zone:* Hardy to Zone 5.

### Buxus microphylla 'John Baldwin'

*History:* An open-pollinated seedling that was found growing in the early 1950s, at the Hoke House in Colonial Williamsburg, Va.

Buxus microphylla *'John Baldwin'*.

*Recommended Plants*

*Description:* With small leaves, it forms a handsome upright plant with straight sides.

*Size:* At 25 years of age, it can be 10' tall and about 3¹/₂' wide.

*Zone:* Hardy to Zone 6.

### *Buxus microphylla* var. *japonica* (Muell.) Rehd. & Wils.

*History:* A native of Asia that was introduced to western culture about 1890. It was often grown around hilltop shrines by the Japanese.

*Description:* Leaves are glossy and nearly round, ¹/₄" to ¹/₂" wide and ³/₄" to 1" long. The new growth, up to three or four years, has twigs that are square in cross section. It flowers freely, which leads to heavy seed production. The numerous and vigorous seedlings, if permitted to grow, will easily overtake and destroy the parent plant. This is the most adaptable of all boxwoods, growing from New England to the Gulf States and California. Plants grown in the sun will have orange or bronze colored foliage during the winter.

*Size:* The Japanese boxwood is an open, broad, quick-growing plant that will reach 8' tall and at least 24' wide. If allowed to layer or set seed, it will easily cover a much larger area.

*Zone:* Hardy to Zone 5.

### *Buxus microphylla* var. *japonica* 'Morris Dwarf'

*History:* Originated as an open-pollinated seedling found by Dr. Henry Skinner at the Morris Arboretum in Philadelphia, Pennsylvania, in 1947.

*Description:* A dwarf boxwood with dense twiggy branches. The clustering of shoots results in an irregularly-shaped top.

*Size:* A 40-year old plant is 3' tall and 4¹/₂' wide.

*Zone:* Hardy to Zone 6.

B. microphylla *var.* japonica *'Morris Dwarf'*.

### *Buxus microphylla* var. *japonica* 'Morris Midget'

*History:* A smaller version of *B. microphylla* var. *japonica* 'Morris Dwarf', with the same origin.

*Description:* Dense and low-mounding with a smooth outline.

*Size:* This very slow-growing

Buxus microphylla *var.* japonica *'Morris Midget'.*

boxwood at 40 years of age can be only 1¹/₂' tall and 3' to 4' wide.

*Zone:* Hardy to Zone 6.

### *Buxus microphylla* var. *japonica* 'National'

*History:* Its origin is the same as *B. microphylla* var. *japonica* 'Morris Dwarf' and 'Morris Midget'. Originally named 'Morris Fastigiate', its habit did not live up to the name, which was changed while Dr. Skinner was at the U.S. National Arboretum.

*Description:* A large, upright plant with strong upright branches and a slightly open habit. Has large, round shiny leaves that stay dark green through the winter.

*Size:* At 35 years it will have achieved a height and width of 15'.

*Zone:* Hardy to Zone 6.

### *Buxus sempervirens* L.

*History:* Native to a broad region which includes southern Europe, North Africa, the coastal region of Asia Minor, the Caucasus mountains, northern Iran, the mountains in Afghanistan and along the Himalayas into parts of China. First described by Linnaeus in 1753. Of the boxwood species in cultivation, *B. sempervirens* is easily the most important. Mature boxwood forests were harvested in Europe for wood to be used in engraving and making tools. This species and its numerous cultivars are, of all boxwoods, the best known by North American gardeners.

*Description:* Grows as a large shrub or small tree.

*Size:* At maturity it can be 25' tall.

*Zone:* Hardy to Zone 5.

*Recommended Plants*

### Buxus sempervirens 'Abilene'

*History:* Originated at Abilene, Kansas.
*Description:* This cultivar has narrow, dark green leaves and an upright habit with a broad base.
*Size:* A 15-year old plant will average 7' tall and 8' wide.
*Zone:* Tolerant to colder climates, it is hardy to Zone 5.

### Buxus sempervirens 'Agram'

*History:* Introduced by the USDA. Grown from seed collected by Dr. Edgar Anderson in 1934 near the Vardar River Valley, Macedonia.
*Description:* A pyramidal and open habit with dark green foliage.
*Size:* A 20-year old plant will be 5' tall and 4' wide.
Zone: Hardy to Zone 5.

### Buxus sempervirens 'Angustifolia'

*History:* This cultivar has been grown since 1756.
*Description:* The treelike habit is similar to 'Arborescens', except the leaves are narrower.
*Size:* A 30-year old plant can be 14' tall and 12' wide.
*Zone:* Hardy to Zone 5.

### Buxus sempervirens 'Arborescens'

*History:* The second most common cultivar used in the landscape. Often incorrectly called "American Boxwood." The common name, "True Tree Boxwood," is a good indication of its mature size. It had been grown prior to 1753.
*Description:* Due to its large size, 'Arborescens' is best suited as a large hedge or in screen plantings. Specimens can live 175 years.
*Size:* Twenty feet tall and 15' wide is rather typical for a mature 40-year old plant.
*Zone:* Hardy to Zone 5 or 6.

### Buxus sempervirens 'Aurea Pendula'

*History:* First described in Europe in 1896.
*Description:* As the cultivar name suggests, this plant has gold variegated foliage with pendulous branches. The variegation is irregular with yellow coloring splashed through the foliage, providing an interesting contrast.
*Size:* A 25-year old plant will average 7' tall and 6' wide.
*Zone:* Hardy to Zone 6.

### *Buxus sempervirens* 'Elegantissima'

*History:* First grown in Europe during the1860s.

*Description:* One of the best variegated boxwoods. The leaves have a bright, irregular creamy-white margin with a green center. Both the leaves and plant are small.

*Size:* The mature size is 7' tall and 7' wide.

*Zone:* Hardy to Zone 6.

### *Buxus sempervirens* 'Fastigiata'

*History:* Collected in 1959 at Hertfordshire, Eng.

*Description:* A unique plant with a very narrow pyramidal habit and clean dark green foliage.

Buxus sempervirens *'Elegantissima'*.

*Size:* A 20-year old plant is 12' tall and 5' wide. The mature height of about 20' is achieved at 35 to 40 years.

*Zone:* Hardy to Zone 5.

### *Buxus sempervirens* 'Graham Blandy'.

*History:* Originally sent to Blandy Experimental Farm, Virginia in the 1930s among a group of boxwoods thought to be from England.

*Description:* This striking plant grows as if it were a trained topiary. The straight sides form a very narrow, upright plant. Spring growth is occasionally pulled down by the spring rains and should be pruned to maintain the parallel sides.

*Size:* A 20-year old plant will be about 9' tall and only 1' to $1^1/_2$' wide. Mature height is 15' to 18'.

*Zone:* Probably hardy to Zone 5.

### *Buxus sempervirens* 'Handsworthiensis'

*History:* Grown in England since 1913, but not widely distributed.

*Description:* One of the strongest-growing cultivars with new growth

***Recommended Plants***

averaging 2¹/₂" to 4". It is a large, dense, multi-stemmed plant, suitable for hedges and screens. Twigs are orange-tinted. The branching has a candelabra effect.

*Size:* This boxwood will easily reach 15' tall with a width of nearly 20'.

*Zone:* Hardy to Zone 5.

### Buxus sempervirens 'Inglis'

*History:* Grown in central Michigan since 1933, it was registered in 1957. Winter hardiness gives this boxwood its popularity.

*Description:* It has a broadly pyramidal habit.

*Size:* Twenty-year old plants will be about 7' tall and 6' wide.

*Zone:* Hardy to Zone 5.

*Buxus sempervirens* 'Graham Blandy'.

### Buxus sempervirens 'Memorial'

*History:* Discovered in a cemetery in Williamsburg, Virginia.

*Description:* Similar to *B. sempervirens* 'Suffruticosa' in appearance. However, 'Memorial' has better winter color, the foliage does not touch the ground, the leaves are slightly longer and the habit is more upright than *B. sempervirens* 'Suffruticosa'.

*Size:* A 25-year old plant is 3¹/₂' tall and 3¹/₂' wide.

*Zone:* Hardy to Zone 5.

### Buxus sempervirens 'Myrtifolia'

*History:* Widely grown in Europe since 1782, but rarely found in North America.

*Description:* A small-sized shrub ideal in today's smaller residential landscapes. The leaves are small and quite narrow, giving this low mounding plant a delicate texture.

*Size:* A 34-year old plant growing under ideal conditions will average 6' tall and 7' to 8' wide.

*Zone:* Hardy to Zone 5/6.

### Buxus sempervirens 'Newport Blue'

History: Grown since 1941, it has quickly become popular due to its
    good winter color. It is originally from Boulevard Nursery in
    Newport, RI.
Description: Blue-green foliage.
Size: It reaches 4' tall and 5' to 6' wide at maturity.
Zone: Hardy to Zone 6.

### Buxus sempervirens 'Pendula'

*History:* As the name implies, this is a boxwood with weeping branches.
    Cultivated since 1869, 'Pendula' has been a constant favorite
    because of its unique shape.
*Description:* It has a unique open habit that gives the plant a J, L or even
    K shape. The branches,
    which have a carefree
    appearance, touch the
    ground and then can layer
    into the ground spurring
    rapid growth.

*Size:* A small plant that at
    30 years of age is only
    5$^1$/$_2$' tall and averages
    about 5' wide if not
    permitted to layer.
*Zone:* Hardy to Zone 5 or 6.

### Buxus sempervirens 'Prostrata'

*History:* Known in England
    since 1914, it has not
    been widely grown in
    North America.
*Description:* This unusual
    boxwood has a low,
    spreading habit similar
    to a ground cover.

Buxus sempervirens *'Pendula'*

*Size:* Rarely grows higher than 3', but can have a width exceeding 10'.
*Zone:* Hardy to Zone 6.

### Buxus sempervirens 'Pyramidalis'

*History:* Grown since 1869, and widely planted.
*Description:* A multi-branched, upright, cone-shaped plant that becomes

***Recommended Plants***

broader as it matures. The branching habit is somewhat open.

*Size:* A pyramidal plant that at 40 years of age is 14' tall and 6' wide. Can achieve a height exceeding 30' with great age.

*Zone:* Hardy to Zone 5.

### *Buxus sempervirens* 'Rotundifolia'

*History:* Grown in Europe for nearly 150 years, and widely available there.

*Description:* The distinctive leaves are nearly round. They average 1" long and ½" wide with no point on the tip of the leaf. The top of the leaf has a convex shape. The foliage is extremely dark green. The plant has a typical shrub shape and is a rapid grower.

*Size:* A 20-year old plant will grow to 10' in height and 9' wide.

*Zone:* Hardy to Zone 5. Exhibits best winter color in Zone 6 and south.

### *Buxus sempervirens* 'Salicifolia'

*History:* Known since 1872, but not widely grown.

*Description:* Called Willowleaf Boxwood due to the similarities of the foliage to the willow. The leaves vary from ¾" to 1¼" long while averaging only ¼" wide.

*Size:* A large boxwood with a single trunk that is likely to reach 12' tall and 15' wide. A 25-year old plant will be about 5' tall and 6' wide.

*Zone:* Hardy to Zone 6.

### *Buxus sempervirens* 'Suffruticosa'

*History:* Its long history extends prior to 1753. Undoubtedly the most popular and widely-grown cultivar of all boxwood. Usually referred to as "English" or True Dwarf boxwood.

*Description:* A rounded plant with tufts of growth resembling a cloud. It has small, rounded leaves giving the plant a dense habit.

*Size:* Noted for its slow growth rate, averaging ¾" to 1¼" per year. There are, however, dwarf cultivars that average only ¼" to ½" per year.

*Zone:* Hardy to Zone 5.

### *Buxus sempervirens* 'Vardar Valley'

*History:* Originated in the Vardar River Valley of Macedonia; collected in 1934 by Dr. Edgar Anderson during his search for cold-hardy boxwoods. Among 155 *B. sempervirens* selected from his collection the best known are: 'Agram', 'Ipek', 'Nish', and 'Vardar Valley'.

*Description:* Retains its dark green color throughout the winter. Spring growth has a prominent bluish cast that slowly weathers off by late summer or fall.

*Recommended Plants*

Buxus sempervirens *'Suffruticosa'*.

*Size:* There are apparently two forms of this cultivar. The most common one is dwarf, growing 3', while the other grows to 7' in height. Both have a broad spreading habit.

*Zone:* Hardy to Zone 4.

**Buxus sinica** (Rehd. & Wils.)M.Cheng **var. *insularis*** (Nakai) M.Cheng (Earlier botanical names were *B. microphylla* var. *koreana* and *B. microphylla* var. *insularis*.)

*History:* Native to Korea and commonly known as Korean Boxwood. Introduced to North America in 1919, this boxwood has provided a large number of superior cold-hardy cultivars. While possessing considerable cold hardiness, the foliage may bronze in the winter.

*Description:* Of generally upright, very loose and open habit enhanced by leaves that are widely spaced and small, $\frac{1}{4}$" to $\frac{3}{4}$" long.

*Size:* The size and shape of this plant can vary greatly but grows to about 6' to 7' tall and wide.

*Zone:* Hardy to Zone 4.

### Buxus sinica var. insularis 'Pincushion'

*History:* Available since 1966, this plant originated at Sheridan Nursery in Georgetown, Ontario, Canada. The name is sometimes incorrectly shortened to 'Cushion'.

*Description:* Appropriately named because the new foliage tips stick out like pins from the very compact foliage.

*Recommended Plants*

*Size:* A 12-year old plant is about 14" high and 20" broad.
*Zone:* Hardy to Zone 4.

### Buxus sinica var. *insularis* 'Tide Hill'

*History:* Originated in western New York in 1932.
*Description:* Has long, narrow, light green leaves. Younger plants have a low spreading habit with a flat top. Older plants tend to have a loose open top.
*Size:* A low plant at twenty years of age will average 18" high and 4' across.
*Zone:* Hardy to Zone 4.

### Buxus sinica var. *insularis* 'Winter Beauty'

*History:* Originated in 1966 at Sheridan Nursery in Georgetown, Ontario, Canada.
*Description:* A round shrub with a compact habit.
*Size:* Mature size is unknown; expected to grow to 3' tall and 4' wide.
*Zone:* Hardy to Zone 4.

### Buxus sinica var. *insularis* 'Wintergreen'

*History:* Originated at Scarff's Nursery, New Carlisle, Ohio in 1960.
*Description:* A popular boxwood with dark green color lasting throughout the winter. Has proven tolerant of the cold winters of Chicago. This cultivar has a medium-sized open habit and is a heavy seed producer.
*Size:* A 16-year old plant will grow to about 5' tall and $3^1/2$' wide.
*Zone:* Hardy to Zone 4.

### Buxus 'Green Mountain'

*History:* Introduced by Sheridan Nursery, in Georgetown, Ontario, Canada in 1966. A dependable cold-hardy boxwood, but not widely available.
Description: Excellent foliage remains dark green throughout the winter. Has a good, dense pyramidal habit. Its parentage is reportedly a cross between *B. sempervirens* 'Suffruticosa' (female) and *B. sinica* var. *insularis* (male), but this is doubtful. There are four putative hybrids, of which 'Green Mountain' is the best.
Size: A ten-year old plant is 36" high and 18" wide; the mature size is unknown.
Zone: Hardy to Zone 4.

# Chapter 3
# Boxwoods in the Landscape

Buxus microphylla *'Green Pillow' along the front edge of the beds and*
B. sempervirens *'Suffruticosa' grown perpendicular to the front of the*
*East Garden at the White House.*

# DESIGN AND RESIDENTIAL USE

Boxwood has been grown successfully in most of the continental states with some exceptions: the Rocky Mountain region, the Northern Great Plains, and those states in Zone 4 and colder. When using boxwoods in the landscape, there are a few design principles that should be considered.

### Enclosure

For living areas or small gardens, taller and faster-growing boxwood cultivars can provide privacy when planted as a hedge. Plants may be allowed to grow together to appear as a solid mass, or spaced sufficiently apart so each is clearly visible.

### Background

Boxwoods may also serve as a backdrop for planting areas. Beds with bulbs, annuals, herbaceous perennials, and ornamental grasses, stand out against the green background.

### Highlights

Boxwoods can frame or highlight flowering borders by grouping them into small masses. The entire border can be "tied together" using low, dwarf forms of boxwoods. Planting boxwoods on either side of a vista can draw attention to a distant view or focal point.

Buxus sempervirens *'Suffruticosa' is used to greet residents and visitors.*

*Landscaping*

Buxus sempervirens *'Suffruticosa'* *is used as a parterre in this rose garden.*

### Intrigue

Great interest is easily obtained in the garden by including a partially hidden area. This encourages the visitor to wander through the garden to see what is beyond the immediately-visible area. Visitors can be pleasantly rewarded by finding a well-planned sitting area that might include a small pool of aquatic plants, a rose garden or any number of other garden features behind a group of boxwoods.

### Design Pattern

The garden with an overall pattern designed to unify and compliment the various elements usually results in a successful garden. Boxwoods are able to provide such a design. Parterres are a good example.

### Year-Round Interest

When the garden is visible from an indoor living area, every effort should be made to plan for interest throughout the year. A well-planned garden can have endless possibilities that provide color or interest all year, especially during the winter months. The attractive fruit of a holly or *Nandina*; the colorful foliage of *Leucothoe*; the interesting bark of the Heritage Birch or the dark green foliage of boxwood are but a few examples of winter interest.

### Outdoor Lighting

Both inexpensive and dependable, outdoor lighting allows the garden to be enjoyed from either inside the house or from the garden. Such lighting can highlight a perennial border, a walking or sitting area, a specimen plant, or any feature that might interest the gardener.

# BOXWOOD GARDENS

There are numerous public gardens where boxwoods are prominently displayed. Here is a brief list of some of the sites in the United States that display boxwood.

### ALABAMA

Arlington Historic Home and Garden, Birmingham. This mansion built in 1842 has six acres of well-landscaped gardens that include boxwood, roses, annuals, and perennials.

### ARKANSAS

Arkansas Territorial Restoration, Little Rock. An entire city block of restored buildings has a boxwood garden and other plants common prior to the Civil War.

*The primary axis of the main peristyle garden at the J. Paul Getty Museum in California. The boxwood is predominantly* Buxus microphylla *var.* japonica *with smaller plantings of* B. sempervirens.

*Landscaping*

## CALIFORNIA

J. Paul Getty Museum, Malibu. There is more than one mile of boxwood hedge among the 12 acres of gardens that surround a replica of the Roman Villa dei Papiri at Herculaneum.

## DELAWARE

Nemours, Wilmington. This is a miniature Versailles with grand fountains, statuary, vistas and a mansion. A sunken garden is edged with *B. sempervirens* 'Suffruticosa'.

The Read House, New Castle. This townhouse, built in 1797, has wonderful old specimens of boxwood.

Woodburn, Dover. Built in 1790, the official residence of the Governor has formal gardens and, best of all, a boxwood maze.

Great Geneva, Dover. This house, built before 1748, has a boxwood and herb garden.

The Homestead, Rehoboth Beach. The boxwood garden, started in 1930, is divided into five sections, including a herb garden.

## GEORGIA

Atlanta Historical Society Grounds, Atlanta. A 23-acre preserve with two restored homes. The Swann House has a formal boxwood garden with both *B. sempervirens* and *B. sempervirens* 'Suffruticosa'.

Founders Memorial Garden, Athens. Three acres of boxwoods in a formal terraced garden including interesting tree specimens.

## KENTUCKY

Liberty Hall, Frankfort. An 1801 home with a well-restored boxwood and rose garden.

## LOUISIANA

Shadows-on-the-Teche, New Iberia. A lovely mansion built in 1834 on the banks of the Bayou Teche. There are old boxwoods and other shrubs as well as wonderful Live Oaks covered with Spanish moss.

Mountain Hope Plantation, Baton Rouge. A restored 1817 home with a boxwood and live-oak garden.

## MARYLAND

Cylburn Wildflower Preserve and Garden Center, Baltimore. On 176 acres, there are extensive wildflower gardens and a good collection of trees and boxwood.

Hampton National Historic Site, Towson. A restored 1783 home and a 48-acre garden with boxwood hedges, formal parterres, and herb garden.

*Landscaping*

Mt. Harmon Plantation, Earleville. A plantation on the banks of the Sassafras River has lots of formally-grown boxwood.

Sherwood Gardens, Baltimore. This seven-acre garden is a place to be in spring, with large quantities of spring bulbs, flowering shrubs, trees, and *B. sempervirens* 'Suffruticosa'.

William Paca House, Annapolis. A well-restored and maintained 18th-century garden has boxwood, roses, and formal and informal parterre terraces.

## MASSACHUSETTS

Longfellow House, Cambridge. At Harvard University, the 1759 home of the famous poet has boxwood hedges with annuals and perennials.

## MICHIGAN

Fernwood Botanical Garden and Nature Center, Niles. This 100-acre facility has prairies, woodlands, wilderness and marsh areas. Daylilies, perennials, a Japanese Garden and boxwoods are just a few of the highlights.

Hidden Lake Gardens, Tipton. There is a wide variety of trees and shrubs, especially boxwoods, in the gardens. A large area for herbaceous plants, including wildflowers, ferns and bog plants, should also be visited. A tropical greenhouse and trial gardens are noteworthy.

## MISSOURI

Missouri Botanical Garden, St. Louis. A world-class botanical garden that has major collections of roses, water lilies, a Japanese Garden, and tropical plants in a half-acre Climatron. The Boxwood Society of the Midwest maintains an excellent boxwood planting here.

## NEW JERSEY

Georgian Court College, Lakewood. This formal garden was built at the turn of the century. It has 16 flower beds all bordered by boxwood. Also, there is a unique Japanese garden.

## NEW YORK

Boscobel Restoration, Garrison-on-Hudson. This early 1800s mansion has a well-restored 36-acre garden with herbs, apple trees, and boxwood. There is an orangery and greenhouse.

Old Westbury Gardens, Westbury, Long Island. The 70-acre garden has individual gardens of boxwood, rose, cottage and Japanese styles.

Philips Manor Hall, Yonkers. Built in the 1700s, this house is surrounded by boxwood hedges and flowering shrubs.

## NORTH CAROLINA

Biltmore House and Gardens, Asheville. Built in 1890, there are 35 acres of formal Italian and English gardens, a large rose garden and a large native plant collection. The shrub garden has boxwood along with many other plants.

Josiah Bell House, Beaufort. The house, built in 1825, has a small garden that includes boxwood.

Reynolda Gardens of Wake Forest University, Winston-Salem. There are 115 acres of gardens here, including four acres of formal gardens with boxwood and flowering trees. There is a wide variety of plant material including trial areas and greenhouses.

Sarah Duke Gardens, Durham. A part of Duke University, this garden has 15 acres of various and attractive gardens.

Tryon Palace Restoration, New Bern. Boxwood is located in the Lathan Memorial Garden among other places. The allée of yaupon, *Ilex vomitoria*, is another attractive feature.

## PENNSYLVANIA

Ambler Campus of Temple University, Ambler. With a College of Horticulture and Landscape Design, there is an extraordinary number of gardens, plants and greenhouses for the students. The ten acres of formal gardens contain boxwood and borders of perennials.

Hershey Rose Garden and Arboretum, Hershey. A magnificent rose garden and a diverse collection of boxwood.

Longwood Gardens, Kennett Square. A world-class garden that should not be missed by anyone interested in gardening at its best. About 125 acres are intensively planted with impressive gardens, excellent collections and beautiful vistas. The magnificent conservatories cover more than four acres. Boxwoods are present in the main fountain garden and other areas.

Morris Arboretum, Philadelphia. Founded in the 1880s, there are 175 acres with magnificent mature collections of numerous trees and shrubs, including boxwoods. There is an excellent, partially-walled rose garden that alone is worth a visit.

Pinchot Institute for Conservation Studies, Milford. This 100-acre site, founded in 1886, is planted with evergreens and boxwood.

## RHODE ISLAND

Green Animals, Portsmouth. A topiary garden begun in 1880, features boxwood and other shrubs pruned into many fine examples. There are several formal gardens that include herbs and perennials.

Shakespeare's Head, Providence. Maintained by the Rhode Island Federation of Garden Clubs, a walled colonial garden has boxwood, herbs, and other points of interest.

## SOUTH CAROLINA

Middleton Place, Charleston. This plantation has grand formal gardens that include *B. sempervirens*.

Rose Hill State Park, Union. Built in 1828, this home has a garden of boxwood planted in the shape of a Confederate flag.

Walnut Grove Plantation, Spartanburg. This plantation, started in the 1700s, has boxwood, herb, and flower gardens.

## TENNESSEE

Craighead-Jackson House, Knoxville. An 1800s mansion with a formal boxwood garden.

James Knox Polk Home, Columbia. This presidential home has a formal garden of perennials framed with boxwood.

Sam Davis House, Smyrna. The restored 1810 house and garden has a good boxwood garden and spring flowering trees, perennials and bulbs.

## VIRGINIA

Agecroft Hall, in Windsor Farms, Richmond. Built in the 15th century, and moved from England; the formal 23 acres display *B. sempervirens* 'Suffruticosa', herbs and ornamental shrubs.

Ash Lawn, Charlottesville. This home of James Monroe has a mature boxwood garden as well as specimens. Herb and cutting gardens should also be visited.

Belmont, Falmouth. Built in 1761, it has boxwood-lined walks with informal plantings of spring flowering shrubs and bulbs.

Berkeley Plantation, Charles City. This home of President Harrison has a boxwood garden along the James River.

Capitol Square Grounds, Richmond. Boxwood surrounds the statue of Stonewall Jackson.

Chatham Gardens, Fredericksburg. Features boxwood, roses, and azaleas.

Colonial Williamsburg, Williamsburg. Boxwood abounds in this beautifully restored 18th-century colonial town. Many houses have private boxwood gardens planted with great historical detail and accuracy.

Gunston Hall, Lorton. Constructed in 1755 for George Mason. The formal two-acre garden, overlooking the Potomac River, has parterres of

*B. sempervirens* 'Suffruticosa'.

Kenmore, Fredericksburg. This mid-18th-century home of George Washington's sister is landscaped with *B. sempervirens* and *B. sempervirens* 'Suffruticosa', and old specimen trees and shrubs.

Mary Washington House and Garden, Fredericksburg. Mature boxwoods are featured in a herb and cutting garden.

Miller/Claytor house, Lynchburg. A boxwood garden is located at the site of this restored 18th-century home.

Montpelier, Orange. This was the home of President James Madison with large old boxwood in the walled garden and extensive *B. sempervirens* hedges along the roads.

Morven Park, Leesburg. This large 1,200-acre estate has a stately boxwood garden and allée, plus a sundial and wildflower garden.

Mount Vernon, Fairfax County. Large boxwood hedges and herb gardens greet visitors as they enter the home of George Washington. Visitors can purchase small boxwood plants which are started as cuttings from the boxwoods growing at Mt. Vernon.

Oatlands Plantation, Leesburg. A grand Federal-style mansion built in 1800 has terraced gardens with boxwoods planted in 1835. There is a wonderful boxwood allée and other ornamental plantings on this 261-acre estate.

Red Hill, Brookneal. A *B. sempervirens* 'Suffruticosa' hedge lines a serpentine walk at Red Hill, the home of Patrick Henry.

River Farm, Alexandria. Home of the American Horticultural Society, the boxwood hedges frame many of the trial and demonstration plantings of daylilies, roses and perennials.

Scotchtown, near Ashland. This is the 1719 home of Patrick Henry.

Stratford Hall Plantation, Stratford. Built in 1725, the birthplace of Robert E. Lee has wonderful boxwoods, herbs, roses, and vegetable gardens.

Sweet Briar College, Amherst. A circle of *B. sempervirens* was planted about 1901.

Virginia State Arboretum, Boyce. There are over 5,000 species of ornamental plants on 100 acres. The American Boxwood Society maintains the Memorial Boxwood Collection. A well-labelled garden, it is one of the most complete collections of boxwood cultivars.

Woodrow Wilson Birthplace, Staunton. Boxwood borders are used in some of the planting beds.

## WASHINGTON, D.C.

Dumbarton Oaks, Georgetown. Designed in 1940, this 16-acre garden has boxwood, an orangerie, and unique formal gardens.

National Cathedral. The Bishop's Garden has a historical planting of *B. sempervirens* 'Suffruticosa', a herb garden and old-time perennials.

Tudor Place, Georgetown. A neoclassical house built in 1805 is surrounded by gardens, lawns, parterres, and woodland which were developed in the Federal period.

U.S. National Arboretum. Established in 1927, gardens and collections include the National Bonsai Collection, National Herb Garden, conifers, Asian plants, azaleas, native plants, perennials and a well-labelled, extensive boxwood collection. There are nearly 140 different cultivars and species of boxwood in this collection.

White House Gardens. Many famous gardens and trees are located on the grounds. There are more than 300 *B. microphylla* 'Green Pillow' in the East and Rose gardens and numerous *B. sempervirens* 'Suffruticosa' in several sites. Garden tours are held on selected weekends in April and October.

*The* Buxus sempervirens *'Suffruticosa' hedge, named Box Circle, is nearly two centuries old. It is in the North Garden at Tudor Place in Washington, D.C.*

*Landscaping*

# SPECIAL USES

Beyond the more traditional landscaping uses of boxwood there are several exciting variations that add a personal and unique quality. These include using boxwood as bonsai, topiary and container plants.

## BONSAI

A *bon* is a tray or shallow container, *sai* means grow. Bonsai means something that is growing in a shallow container, and is the art of growing a living tree in a small pot. There are three training methods in which the plant is dwarfed. The Chinese method, known as Penjing, was developed first. Later the Japanese method of bonsai began. Finally, North Americans have developed a third style. Each is unique in its own way. Since it is an art form, the styles are constantly evolving.

While nearly any plant will make a suitable bonsai with proper training, azalea, beech, ginkgo, maple, oak, and pine are favored for use as bonsai. *B. microphylla* 'Compacta' and *B. microphylla* var. *japonica* 'Morris Dwarf' are also used in the art of bonsai.

*Beginning bonsai enthusiasts often choose* B. microphylla *'Compacta' because of its miniature tree-like characteristics and its ability to adapt easily to the demanding cultural conditions placed upon bonsai plants.*

***Landscaping***

Unfortunately this is not the proper place to give any details on the fascinating procedures to grow these wonderful miniature replicas of trees. However, there is an exceptional number of books available on the subject. Quite often public libraries and some garden centers will have introductory books on bonsai.

## PARTERRE

The French, during the Renaissance, placed great importance on the art of gardening. Gardeners began to compete with each other, often resulting in large majestic gardens which required a vast labor force. Trees planted closely together then clipped uniformly and topiary became a dominant theme. This technique developed into the parterre, which is flower beds with paths of gravel or turf arranged in various geometric patterns. The parterre was fully developed by French gardeners and flourished during the sixteenth and seventeenth centuries. *Buxus*

Buxus sempervirens *'Suffruticosa', grown as a hedge to form a parterre, is interplanted with* Buxus sempervirens.

*sempervirens* and *B. sempervirens* 'Suffruticosa' were extensively used for edging and for high hedges between areas. B. *sempervirens* 'Suffruticosa' is often referred to as "Edging Box" because of its extensive use as a border plant in parterre gardens.

## TOPIARY

Topiary is a formally-pruned plant in a particular shape such as an animal, ball, or a glass-top-smooth hedge. Topiaries have been described as "...characterized by the clipping or trimming of live shrubs or trees into decorative shapes, as those of animals or birds."

Romans began the art of topiary. They were fond of topiaries and had gardeners whose sole task was to create and maintain topiaries in the gardens. In the United States the mid-1950s through the mid-1970s saw an interest in topiary. One of the best examples is the Ladew Topiary Gardens in Monkton, Maryland.

A few of the plants used to create topiary include: Yew, *Taxus*; Hemlock, *Tsuga*; Creeping Euonymus, *Euonymus radicans minimus*; Privet, *Ligustrum*; and Boxwood, *B. sempervirens*. The order given also corresponds to the adaptive nature of the plant to topiary. Yew is the best plant while boxwood is among the least successful plants for topiary, due to the costant shearing that slowly weakens boxwood.

In managing topiary boxwood, the plants should be sheared in early June. To produce a boxwood hedge, the base should be slightly wider than the top. For

Buxus sempervirens *trained as a topiary at Colonial Williamsburg, Va.*

shearing techniques, see Shearing in the pruning section, p. 48.

## CONTAINER USE

### Which boxwoods will grow in containers?

A container-grown boxwood provides unique interest in a garden.

*Landscaping*

Depending on the desired effect, either a dwarf or medium-sized box-wood can be placed in a container. Suitable dwarf boxwoods include: *B. microphylla* 'Compacta', 'Grace Hendrick Phillips' and 'Green Pillow'; *B. microphylla* var. *japonica* 'Morris Midget'. Dwarf boxwoods have an advantage because they do not require pruning in order to keep a proper scale with the container. These plants naturally grow slowly, stay small and possess thick full branchings.

There are many medium sized boxwoods suitable for containers; these include: *B. microphylla* 'John Baldwin', B. *micro-phylla* var. *japonica* 'Morris Dwarf', *B. sempervirens* 'Graham Blandy', *B. sinica* var. *insularis* 'Justin Brouwers' and 'Tide Hill'.

### What is the best container?

In residential situations, there are numerous types of attractive containers to select from. Materials include redwood, oak, fir, cedar, cypress, cast stone and clay. Depending on box-wood selection, the dimensions of the container may vary. Consider using more than one container, or different sizes and types of containers which will highlight the individual charac-teristics of the box-wood. If the container is to be constructed of wood, there should be 8 to 12 drain holes ³/₄" in diameter, in the bottom panel. The container should have

Buxus sempervirens *grown as a standard in a clay container.*

1" to 2" legs at the bottom of each corner to allow for water drainage and provide air circulation.

### How is the container planted?

Early spring or mid-fall is the preferred season to plant the container. The soil should be a light, loamy soil mixture with generous portions of humus. Use enough soil so the top of the rootball will be 3" to 4" below the top of the container. Remember that the soil will settle slightly after planting. To compensate for this, the boxwood should be planted slightly higher than the final, desired level. Do not place coarse gravel or rock in the bottom of the container. Gravel in the bottom actually reduces both the area for root growth and drainage.

Before the boxwood is placed in the new container, remove any plastic container or synthetic burlap. Natural burlap can remain around the rootball if loosened at the top. Check the rootball for exposed roots at the edge of the ball. If many are visible, use a soil knife and cut the roots and soil ball in several places. This is necessary to permit the roots to grow into the new soil in the container. After the boxwood has been placed in the container, add soil around the rootball, then water generously. Finally, add 2" mulch to conserve moisture and to reduce soil temperature fluctuations.

### Where can I put the containerized boxwood?

Containerized specimens are best placed in a site that has some shade, away from strong winds, water dripping from the eaves of the house and where the container will not sit in water. Ideally, the container should be periodically rotated by a half turn to equalize the exposure and keep the plant healthy on all sides. Certainly an important consideration is placing the container where it can be easily viewed and the boxwood can be fully appreciated.

### What about watering a container plant?

Boxwoods, or any plants for that matter, placed in a container are subject to a higher level of cultural stress than a similar plant in the ground. Beginning in early spring and continuing through late fall the containerized boxwood needs regular watering. During the hot weather, it may be necessary to water it daily.

There is no firm guideline for the actual amount of water necessary, other than checking the appearance of the soil and the plant itself. While watering, spray a fine, hard mist of water through the foliage of the plant to wash off dust, cobwebs, and to reduce the red spider mites. Early morning is the best time to water.

## What happens in winter?

Boxwoods left in wooden tubs will do fine outside in Zone 8 and areas warmer. Plants in cast stone or clay pots in a freezing climate need protection to keep the container from cracking. Putting the container into a trench with straw or leaves packed around will provide sufficient protection for boxwoods growing in Zone 5 to 7.

# Chapter 4
# Culture of Boxwood

*Bronze or orange discoloration is common on stressed boxwood in the spring and in the fall. The stress is caused by poor cultural conditions. As cultural conditions decline, the discoloration becomes more intense.*

# CULTURE OF BOXWOOD

Boxwood is a relatively low-maintenance, durable, ornamental shrub. The popularity of boxwood is due, in part, to its low-maintenance requirements. Trouble usually begins when a gardener reads the phrase "low-maintenance plant," because it is often considered equivalent to "no-maintenance plant." All plants, even those that require little care, should be observed regularly. Problems or changes noticed at an early stage can help prevent major problems in the long run.

## SEASONAL GARDENER

A change in seasons also means a change in garden tactics. Certain tasks are best performed at certain times of the year to permit the boxwood to flourish. Listed below is a gardener's guide to boxwood cultural tasks and the best season to practice them.

| | | |
|---|---|---|
| **Spring** | Monitor Insects | (see p. 62) |
| | Sanitation | (see p. 49) |
| | Transplanting | (see p. 44) |
| | Watering | (see p. 49) |
| | Weed Control/Mulch | (see p. 42) |
| | Repair winter injury | (see p. 51) |
| **Summer** | Monitor Insects | (see p. 62) |
| | Watering | (see p. 49) |
| | Weed Control/Mulch | (see p. 42) |
| **Fall** | Fertilizing | (see p. 39) |
| | Mulching | (see p. 42) |
| | Planting | (see p. 44) |
| | Pruning | (see p. 47) |
| | Transplanting | (see p. 44) |
| | Watering | (see p. 49) |
| | Weed Control/Mulch | (see p. 42) |
| **Winter** | Fertilizing | (see p. 39) |
| | Protection | (see p. 53) |
| | Pruning | (see p. 47) |
| | Thinning | (see p. 47) |
| | Repair winter injury | (see p. 51) |

*Culture*

## ABIOTIC CONDITIONS

Boxwoods, like most living organisms, are subject to stress. A plant can be stressed by biotic conditions such as insects, diseases and weeds. However, boxwoods can also be stressed by abiotic conditions. A wide variety of outside forces will affect the quality of boxwoods. Of all of these, water and cold stress have the greatest effect on determining where and how well a boxwood will grow.

### Air quality

Smog can be particularly acute in large cities during the summer. Poor air quality will adversely affect boxwoods. Levels of $O_3$ (ozone) at 0.02 to 0.1 ppm for one hour per day can produce plant damage. As an example, Los Angeles averages 137 days per year with an $O_3$ concentration above 0.12 with levels reach up to 0.33 ppm.

### Cold

Foliage bronzing, which typically occurs in late fall and early spring, usually indicates a culturally-stressed plant. There are many causes and the individual site must be evaluated to determine a correct treatment.

### Excess fruiting

Heavy fruiting can be an indication of stress. There are many causes, and the plant needs to be examined for corrective measures.

### Leaf drop

Some leaf drop is normal and expected. If the plant doesn't maintain three years' worth of foliage, look for stress conditions.

### Maintenance practices

Mowers can damage branches and cut the foliage. Improper mulching is a serious concern (see p. 41, the section on mulching). Over or under watering will affect the health of the plant (see p. 49, the section on water).

### Soil compaction

This is a concern for home gardeners where new home construction occurs, or where heavy equipment has been used, prior to planting. The compressed soil stunts root growth by reducing drainage, water penetration, and soil air content.

**Restriction of root zone**

Boxwood roots grow wide and shallow. When planted next to a barrier such as a foundation or sidewalk, the optimum range of root development is reduced, resulting in a weak, stunted plant.

**Salt**

Boxwoods are salt-sensitive plants. Salt pulls moisture out of the plant, creating a water-stressed plant. Salt also accumulates at the tip and edge of the leaf, leaving a distinct line between live and dead tissue.

A soil test can determine if the salt level is high enough to cause a problem. Boxwoods thrive with an electrical conductivity (EC) less than 0.75 decisiemens per meter (dS/m). They can not tolerate salt concentrations in the soil above 2.0 dS/m.

A concern for boxwood planted near a road in areas that experience snowfall is road de-icing salts. Sodium chloride is a commonly used material that is toxic to boxwood. Potassium chloride, potassium sulfate, or fertilizer, are preferable because they are less toxic. High salt concentration can be corrected by periodically flushing the soil to leach the salts out of the soil.

**Excessive sunlight**

This can cause sunscald to the leaves and contributes to bark winter injury. It is also responsible for *B. sempervirens* 'Suffruticosa' foliage growing overly thick.

**When you're unsure of what's wrong**

There are a few questions to ask yourself to determine the cause of leaves that are discoloring or parts of the plant that are dying:

| | |
|---|---|
| Is the soil pH favorable to boxwood? | (see p. 40) |
| Is there adequate drainage? | (see p. 49) |
| Is the site too shady or too sunny? | (see p. 41) |
| Is there turfgrass or tree root competition? | (see p. 41) |
| Were the boxwoods recently mulched? | (see p. 42) |
| Has the foliage been thinned? | (see p. 47) |
| Is there a buildup of debris in the center of the plant? | (see p. 49) |
| Have the boxwoods been receiving adequate water during the past year? | (see p. 49) |

*Six of the essential elements are obtained from the soil in relatively large amounts. This plant illustrates a deficiency in two of them. A lack of magnesium causes the orange or gold tip, and lack of phosphorous causes the tan-colored margin. Maintaining a pH of 6.5 to 7.2 insures the maximum availability of all six nutrients.*

## FERTILIZING

### How do I know whether I should fertilize my boxwoods?

There is no regular schedule to guide fertilization of boxwoods. The most reliable guide to applying fertilizer is by testing the soil. Soil samples analysis by the Cooperative Extension Service will provide appropriate fertilizer recommendations for a specific site.

If the boxwood begins to show symptoms of nitrogen deficiency, then it may be time to fertilize. The earliest symptom of nitrogen deficiency is yellowing of lower leaves. It will have a rather uniform yellowing, that is more pronounced on the older leaves inside the plant. The leaves then become smaller and thinner and turn quite bronze in winter. Boxwood leaves will normally stay on the plant for three years. If they fall off earlier, this may be a symptom of nitrogen deficiency.

### What kind of fertilizer is best and when should I use it?

Phosphorus (P) and potassium (K) are seldom lacking for boxwood in most soils. A common fertilizer such as a 10-6-4 analysis contains 10% nitrogen, 6% phosphorus, 4% potassium and possibly other

*Culture*

nutrients. This complete or balanced fertilizer creates excessive amounts of phosphorus and potassium in the soil resulting in waste, salt buildup and contributes to water pollution. Nitrogen is the nutrient to which boxwood and other plants most commonly respond.

So, if it has been determined through soil testing that nitrogen is lacking, late fall applications of a controlled release fertilizer promote root growth and provide best results. The fertlizer should have something close to a 3:1:1 ratio, for example, a 17-6-10 analysis would be appropriate. The amount of fertilizer to use, or the application rate, would be determined by the soil test.

### What is the best method to apply fertilizers?

Broadcast fertilizer around the base of the plant, just beyond the drip line. Surface application is the easiest and fastest technique. It is effective around the drip line because the most active roots are located there. Fertilizer particles that come into direct contact with the roots of unmulched boxwoods can cause root burn. If the fertilizer is over-applied, this will cause the foliage to brown and may even result in branches dying. This can be avoided by broadcasting fertilizers only on mulched boxwoods when the soil has adequate moisture.

Deep root fertilization, drilling holes and filling them partially with fertilizer, is not recommended. While it does eliminate volatilization of urea and ammonium, it is not worth the effort. The roots of boxwood grow close to the surface and they do not benefit from deep root fertilization.

### What does soil pH have to do with fertilizer?

The pH needs to be in a proper range in order for the nutrients to be available to the plant. The optimum soil pH for boxwoods is between 6.5 and 7.2.

If the pH is below the recommended range, add dolomitic lime. This lime has a low oxide content and will persist in the soil for 4 to 7 years depending on application rates and soil type. If the pH is above 7.5, then lowering the pH would be in order. This is accomplished by adding iron sulfate. Your County Extension Service is an excellent resource for recommendations for properly adjusting soil pH. A soil test probably will be needed.

## SITE SELECTION

Location is very important when planting boxwoods. Healthy boxwoods planted in a poor site will perform poorly or may die. Each individual site will have its own set of conditions. The following is a list of site traits to consider.

## Acidity

Boxwoods perform best at a pH of 6.5 to 7.2.

## Drainage

See water section p. 49.

## Exposure

Boxwoods do best if they have partial sun during the growing season. However, during winter the site should offer protection from sunshine and wind. Plants exposed to continual, direct sun in winter will have reddish-brown or yellow leaves due to rapid temperature changes. Windy sites in winter will cause the boxwood to lose water faster than it can be absorbed. This will also cause reddish-brown discoloration of the leaves. Boxwoods planted close to the south or west sides of buildings often experience winter bronzing.

## Soil Texture

A loamy soil which has nearly equal portions of sand, silt and clay is ideal. A sandy soil generally does not have sufficient water-holding capacity. Heavy clay soils often lack good drainage which results in roots being constantly wet, reducing available air. When the clay soil does dry out, it is quite hard, inhibiting root growth. To correct heavy clay soil, first add organic matter to create pore space for proper root development; then install a tile system before planting to ensure good drainage.

## Root Competition

Competition for water and nutrients from nearby trees, shrubs, and turfgrass should be avoided. Ideally, the site should be free of root competition. The roots of many plants will reach out as far as the drip line, some much farther. Site conditions can be modified by appropriately increasing water and fertilizer applications to compensate for the loss due to root competition. An effective long term solution involves transplanting the boxwood or the interfering plant.

The majority of boxwood roots are near the soil surface. Because of the extensive lateral root system, boxwoods should not be planted too deeply or in locations that restrict the spread of the roots. Root development will be stunted if growing too near the foundation of a house, sidewalk or other barrier.

## MULCH

### Why should I mulch?

It conserves soil moisture by reducing evaporation from the soil surface and reduces weed growth. Weeds will use the water that would overwise be available to the boxwood. Mulch aids water penetration into the soil, slows the full force of heavy rains, and reduces the danger of soil erosion. By slowing down the movement of the water, it is more likely to be absorbed by the soil than to run off.

Soil temperature is moderated since the mulch, acting as an insulator, reduces temperature fluctuations. The soil remains cooler in the summer, which encourages root growth. In the winter the soil is not as likely to alternately freeze and thaw, which may injure the roots near the surface.

Mulch greatly inhibits weed growth. Weeds, particularly vines, will shade the boxwood, and use the water and nutrients in the soil which would otherwise be available to the boxwood.

Certain mulches are attractive to the overall appearance of the garden.

Soil fertility is increased as nutrients leach from the mulch and as the mulch decomposes.

### Which mulch is best?

#### Shredded hardwood bark
Because of its availability, this mulch is the most popular in the northern states. This is an effective and visually pleasing mulch that also retains soil moisture and inhibits weed growth.

#### Pine bark
Relatively inexpensive and readily available, this mulch is often chosen for gardens in the gulf states. The medium and fine grades of this mulch, however, tend to blow away with the wind and wash away with rain while coarse grades of pine bark do little to discourage weed growth.

#### Pine needles
This is a very effective and attractive mulch. Unfortunately it is commercially available on only a limited basis.

#### Hardwood chips
Professional tree companies grind up the branches of trees in a chipper. This by-product then becomes a useful coarse mulch. It is preferred to allow fresh wood chips to age and begin to decompose before using it in garden beds. Fresh material removes a great deal of nitrogen during the decomposition process and should not be placed

directly into planting areas. The chips also release manganese to the soil. If the chips are applied too heavily, they will release an excess of manganese which is toxic to plants.

### Geotextile fabric

Also known as weed barrier cloth, this fabric is applied under mulch. It gives a long-lasting and effective method of halting weed growth. The material permits water to pass through and when it is placed on a slope, mulch is not as likely to slide downhill as compared to a plastic barrier.

### Compost and shredded leaves

Decomposed leaves or compost from the home is an excellent source of nutrients and organic material when applied lightly over the soil.

## Some materials are not desirable as mulch

### Sawdust

Available in limited areas, sawdust should only be used after it has become well decomposed. If fresh sawdust is applied to the garden, it will remove large amounts of nitrogen from the soil as it decomposes. Pressure-treated sawdust, which contains toxic wood perservatives, should not be used. Black walnut sawdust should not be used due to its allelopathic effects. Sawdust has a tendency to cake and become hard, leading to the shedding of water.

### Peat moss

Shredded peat moss is preferred as it does not compact, but in general is more effective when lightly cultivated into the soil than used as a mulch. It is beneficial in soils that have less than 1% organic matter. However it has disadvantages and will reduce soil aeration, lower the soil pH, harden, and shed water.

### Stone chips and gravel

A durable, permanent mulch that is available in a wide range of materials, sizes, shapes and colors providing an effect not possible with organic mulches. Used by itself, however, stone is not effective in controlling weeds or improving the soil. Marble gravel is quite alkaline and should not be used.

### Black plastic

Black plastic is more durable than clear or white plastic. While effectively stopping weed growth, it does not allow air or water to pass through and severely impairs normal water movement in the soil. If used on even a slight slope, mulches placed on top of the plastic tend to slide downhill.

**What is the best way to apply mulch?**

Before applying any mulch make sure that there is a good supply of soil moisture. Apply mulch after several hard frosts in late fall. Mulch may also be used in early spring after the chance of hard frosts has passed. Apply mulch only to a depth of 1". If the mulch is being used to top dress existing mulch, remove some of the existing mulch and apply only enough to bring the total depth to 1". Avoid mounding mulch under branches, which will encourage adventitious rooting. Over a period of time the organic materials used in mulch will decompose adding humus and plant nutrients to the soil. As this occurs, it will be necessary to add additional mulch to maintain the 1" mulch depth.

**Can mulching cause problems?**

**Nitrogen deficiency in the soil**

By using freshly chipped wood or bark mulch, including sawdust and straw, the soil nitrogen is rapidly used by the mulch as it decomposes, leaving little for the surface-rooted boxwoods. This is easily avoided by using only decomposed mulch. If fresh mulch has already been applied, high nitrogen fertilizer can be applied by broadcasting.

**Excessive moisture**

This is a problem in poorly drained soils. Even in well-drained soils the lower average temperature and higher moisture content make less nitrogen available. A remedy for this is to discontinue mulching and consider improving the water drainage.

**Excessive mulch**

It is difficult to overstate the problems associated with applying too much mulch. A similar situation can occur when repeated applications of mulch are permitted to accumulate. Mulch that is deeper than 1" will encourage voles. The proper growth of the root system is stunted. It will also encourage the bottom boxwood branches to layer.

There are soil problems associated with high organic matter content that mulch will exacerbate. Further, the roots under excessive mulch are denied the proper levels of air and water.

## PLANTING AND TRANSPLANTING

When planting a new boxwood, it is important to know how large it will be at maturity. Future problems of overcrowding can be avoided if the ultimate size is known. This allows the plant to be spaced properly when initially planted. Small- to medium-size boxwood can be successfully transplanted by the homeowner from one area to another. To move large boxwood, it is best to seek help from a professional landscape contractor.

*Initial cut made to form the trench. Branches are covered with burlap for protection.*

*Base of root ball is undercut at a 45° angle.*

*Drum lacing the root ball with sisal rope.*

*Chain placed around the root ball to facilitate the move.*

**When is the best time to move boxwoods?**

Planting and transplanting is best done in early fall, generally October. This allows the boxwood time to produce new roots which will help the plant take up water during the winter. As long as the ground is not frozen, the plant spends the majority of its time in winter producing new roots.

An alternative season is early spring, before new growth starts. Late spring and summer can be a difficult time for moving boxwoods. Monitoring cultural conditions is important in this case.

**How are boxwoods moved?**

Boxwood roots seldom grow very deep in relation to the total plant. The roots will, however, spread a great deal, so in digging the root ball,

*Culture*

width is more critical then depth. The root ball ought to be at least as wide as the drip line. The depth of the ball is usually determined by the height of the plant. A 3:1 ratio provides a general guideline. For example, a 6' tall boxwood should have a root ball 1½' to 2' deep. Root prune larger plants a year in advance if practical. To facilitate digging and transporting, tie up the branches of medium to large plants with rope or heavy twine. Digging should not be attempted if the soil is excessively dry or wet. If too dry, water the boxwood thoroughly several days in advance.

Before digging, mark a circle on the ground at the approximate "drip line" of the foliage. The majority of the feeder, or hair roots are in this zone. This line, at minimum, is where the outside of the root ball will be. Using a flat-edge spade with its back turned to the plant, cut straight down 9" to 12". Continue all the way around the circle. Dig a trench outside the circle. Then trim the root ball by removing surplus soil so that it is uniformly tapered. Be sensitive to the amount of roots being cut by the spade. If a large number of roots are being cut, it would be prudent to ease off cutting at that area and make the ball wider. Conversely, if no roots are found, the ball can be dug closer to the trunk. After trimming, undercut the root ball at an angle of 45°.

With small root balls, those under 10" to 12" in diameter, the plant can be removed from the hole using two spades inserted under the ball from opposing sides. For slightly larger root balls, 8" to 15", it will be necessary to tip the root ball to one side and insert a burlap bag under the ball. Then lift the plant by grasping the corners of the burlap. For root balls larger than about 16", the root ball should be wrapped in burlap and then reinforced with rope. This will keep the root ball from falling apart.

### How are boxwoods planted?

At the site, dig the depth of the hole the approximate size of the root ball. It is best to place boxwoods on firm soil. One of the most common mistakes in transplanting is to plant a boxwood too deeply. Even planting a boxwood at the existing soil line is too deep. The top ⅛ of the root ball ought to be above the existing soil level. Place the ball into the hole. If synthetic burlap is used, remove it from the ball at this time. Natural burlaps can remain on the ball and only need to be loosened from around the trunk. Fill the hole with soil and water slowly at the base of the plant.

Do not thin the foliage after planting or transplanting. It is the leaves that produce sugars the plant needs for root growth. By reducing the leaves, the plant's ability to produce new roots is proportionally reduced. The more vigorous the root growth, the quicker the plant becomes established. Desiccation after planting is a concern which is discussed in both the water and winter effect sections see pp. 49, 51.

## PRUNING

Slow-growing, broadleaved, evergreen shrubs, like boxwoods, have the majority of their growth from buds at or near the ends of the branches. Some develop a dense outer shell of foliage with only a few leaves in the center. Most require little pruning except when an occasional branch outgrows the general habit of the plant. Boxwoods develop more of a main-branch framework than do many faster growing shrubs (i.e., *Forsythia*) that tend to sprout vigorously from the base.

Some boxwoods have central leaders with tall, straight trunks such as *B. sempervirens* 'Fastigiata', 'Pyramidalis' and 'Graham Blandy'. Others have several main branches with spreading crowns such as *B. sempervirens* 'Suffruticosa' and *B. microphylla* var. *japonica*. Then there is the low-mounding habit which would include *B. microphylla* 'Compacta', 'Green Pillow' and *B. microphylla* var. *japonica* 'Morris Midget'. Between these extremes, many intermediate forms occur. Proper pruning can exploit the natural characteristics of a specific cultivar as well as enhance its appearance and maintain its health. There are three different methods; each may be appropriate depending on the desired results. They are thinning, shearing and heading cuts.

## THINNING

Thinning is a type of pruning that reduces the number of branches at the outer edge of the shrub. It removes foliage from around the outside of the boxwood, making the foliage less dense. This is a simple procedure. For example, on a typical 5'-tall *B. sempervirens* 'Suffruticosa', many small branches 6" to about 8" are pruned out in a random fashion. Continue thinning various branches until the interior branches are slightly visible. Preferably, this should be done annually, but is necessary at least every two years. Thinning can be done anytime the weather is suitable for working outdoors, when the temperature is above freezing. Boxwood branches are very brittle and can break easily in freezing weather. The preferred time to do the thinning is late November to December. The cuttings can be saved and used for holiday decorations.

*B. sempervirens* 'Suffruticosa', the so-called "English" boxwood, requires the most attention for thinning. A few other boxwoods require periodic thinning: *B. microphylla* 'Compacta', 'Green Pillow', *B. microphylla* var. *japonica* 'Morris Dwarf' and 'Morris Midget', *B. sempervirens* 'Memorial' and 'West Ridgeway'.

Thinning is the single most important maintenance activity for keeping a *B. sempervirens* 'Suffruticosa' healthy. Without adequate light or air circulation, the interior leaf shoots die, leaving very little green foliage on the majority of the branch. This is due to overly thick foliage at the ends of the branches. Thinning will allow the center of the plant to receive the proper amount of sun and air. Under these conditions the boxwood will have leaves growing on the entire length of the branch. Much of the poor health of boxwood is a result of not thinning the plant. For example, over-thick foliage encourages *Macrophoma* leaf spot and *Volutella* leaf and twig blight, see p. 75-78. It will also allow the branches to become thin and weak making them more susceptible to breakage by snow and ice.

## SHEARING

Boxwoods are occasionally sheared to control the size or the shape. Topiary is an example of shearing. While the effect is usually quite spectacular, it is not their natural growth habit. Electric shears are best used to achieve straight edges for a hedge. Hand shears will give best results when used on topiary.

The best time to shear the plants is in early June. A second light pruning may be desirable in July to keep the foliage neat. It is best to complete all shearing before August, or by July 15 in Zones 6 and colder. Pruning after these dates can encourage late growth that may still be tender when freezing weather approaches.

Boxwoods are highly stressed by shearing. Shearing encourages dense, multiple tip branching, leaves are cut in half, and diseases are more prevalent. However, once these concerns are understood, they can be compensated for by proper care. For example, in addition to shearing to maintain the desired shape, the boxwood must also be thinned. Thinning the plants permits the inner portion to receive light and air which will encourage growth on the inner stems and reduce the potential for diseases.

## HEADING CUTS

There are two reasons for heading cuts: to control growth, or to restore overgrown boxwoods. As the boxwoods mature, they begin to take up more space than what was probably first planned for. Reducing the size of overgrown boxwoods requires drastic corrective pruning. This work is best done in early winter. After a major pruning, the boxwood may appear unsightly until the foliage fills in. Planning ahead by proper pruning can often avoid this.

If a small amount is to be cut the entire pruning can be done at one time. If a more severe pruning is needed, then a two-step approach is best. First cut large branches on just one side of the plant. The second year, the other half of the plant would be pruned in the same manner. To make the cuts on large branches, a small, sharp-toothed, curved hand saw will perform the best. The cuts should not be made flush against the crotch, but just beyond the natural collar. This will promote the most rapid healing and compartmentalization of the wound area.

There are two concerns associated with pruning cuts. Fresh pruning cuts attract female boring insects looking for a place to deposit eggs. These cuts are also open to infection if made during wet or moist weather when canker-forming fungi are sporulating. Tree coats should not be applied to pruning cuts because they may actually injure the plant and delay healing.

## SANITATION

Maintaining a culturally clean plant is important for all boxwoods, especially *B. sempervirens* 'Suffruticosa'. The incidence of disease, pest damage and abnormal growth can be avoided by proper sanitation techniques.

Once a year the leaves and twigs that have accumulated in the center of the boxwood ought to be cleaned out. A thinning, shake, then rake strategy gives the best results. First, thin the plant to reduce the dense foliage and prune out dead twigs. Then, a vigorous shaking of the branches will force the debris to fall to the ground. Finally, a leaf rake can be used to collect the debris which should then be removed from the site.

A build-up of debris will encourage aerial root growth on the branches. These adventitious roots, which are not protected, are easily damaged during periods of dry or extremely cold weather. When the crown is kept clean and free of dead leaves and twigs, the plant will maintain proper air and sunlight circulation. Debris inside the shrub will catch the snow which pulls down and breaks the branches. Periodic cleaning improves the air and sun circulation through the interior of the plant, this reduces the chance for infection by diseases such as *Macrophoma* and *Volutella*.

## WATER

Maintaining proper moisture in the soil is the single greatest factor for landscape plants. Water is vital to the well-being of a plant, and equally important is the gardener who applies the water. Let's start from the ground up.

### Water in the soil

An ideal soil is roughly 45% minerals, 25% air, 25% water, and 5%

organic matter. An important element in maintaining the proper amount of water in the soil is drainage. For example, a sandy soil will permit water to move quickly through the soil. Conversely, a clay soil will hold onto the water, causing a poor drainage condition. The roots of boxwood need a continuous and adequate supply of oxygen to grow properly.

Too much water in the soil will reduce the available oxygen. Conditions such as a compacted soil, high water table, or flooding will interfere with the oxygen that boxwood roots need. To determine if the soil has proper internal drainage, dig a hole one foot in diameter and a foot deep. Fill the hole with water. If the water is not gone within one hour, the site is not well drained.

## Water in the plant

In a dry period, the shallow boxwood roots can not extract water from the soil as transpiration continues. This draws water out of the cells and the plant wilts. A somewhat similar situation occurs in winter. The water in the soil has frozen, and is then unavailable to the plant. If the drought or winter is too severe, the plant may lose too much water. Under this condition a permanent wilt occurs and a branch or even the entire plant dies. If these extreme conditions are temporary, the plant can re-hydrate by recovering sufficient water.

## Water in the atmosphere

For most plants, 95% to 97% of all water taken up by the roots is lost to the atmosphere through the stomates on the underside of the leaves. This process is called transpiration. An additional 1% to 3% is lost through the cuticle layer of the leaf.

## Water quality

The Midwest region has ground water with boron levels high enough to be toxic to many plants, including boxwoods. Levels exceeding 0.5 ppm (parts per million) can cause injury that is similar in appearance to salt damage. Some areas experience water with a high sodium, or salt, level. Water with an electrical conductivity (EC) less than 0.75 decisiemens per meter (dS/m) is best for boxwoods.

## Watering the boxwoods

The best guide to watering is by watching the soil. The soil should be slightly moist from the surface to 12" to 18" below the surface all year round. To check for this, begin by digging several holes 12" deep. Make these holes in various spots throughout the garden at different times of the year. Then one can see how far down into the soil the moisture is penetrating. Only through time and observation can an appropriate watering schedule be developed for a particular site. There are still other variables that will affect watering. Some of them include: the type of soil, the size of the plant, amount of rainfall, air temperature and relative humidity, slope of the soil, wind, sunlight, type and thickness of mulch, and how recently the plant was moved.

As a general guide to watering, boxwoods need about 1" of rainfall every 10 days from spring to fall. If the fall weather is dry, water thoroughly just before the first heavy freezing weather is expected. If dry weather continues into the winter, water the equivalent of 1" of rainfall every three weeks whenever the ground is not frozen. There are numerous variables to consider when using this general watering guide. Refer to conditions in the previous paragraph.

Preferred watering devices include soaker hoses and trickle irrigation. It is important to water to a depth of about 18". This will encourage the roots to grow deep into the soil. Overhead watering only wets the foliage which encourages *Volutella* leaf and twig blight, see p. 77. Additionally, 15% to 20% of the water can evaporate before it is absorbed into the soil.

## WINTER EFFECTS

Cold winter temperatures are a major influence on where boxwoods can be grown. Boxwoods with a southern exposure, or with a winter temperature colder than 20°F, are more predisposed to winter damage. The effects of winter can be minimized by sound cultural practices as well as proper plant and site selection.

The branches of *Buxus sempervirens* are generally hardy to -5°F, USDA Plant Hardiness Zone 6. A few cultivars are hardy to -20°F, Zone 5. It is on this basis that plants are rated for hardiness. While the roots are more sensitive to cold, generally surviving only to 15°F, they are protected from cold injury by soil and mulch.

### What does winter damage look like?

**Discolored foliage** may be reddish brown, yellowish, grayish green or in extreme cases have a complete loss of color. The ends of a branch are the most prone to winter effects.

**Sunken bark** along the trunk just above the ground or in the crotches and along the sides of the main branches. Close examination of the sunken bark may reveal that it is brown throughout or contains brown streaks.

**Removed bark** can occur on the twigs and branches. The bark will separate from the wood, patches of considerable size can be stripped off. In severe cases, the stem will crack and death of entire branches can occur.

**Cracked bark** is a type of winter damage which may not become evident until the middle of the summer. This cracking is caused by the weight of snow or ice. Frequently the snow causes a hairline split in the bark and vascular cambium. The branch will stay green through the spring. As the heat of summer approaches the vascular cambium is not able to meet the increased water demands of the leaves, and the branch dies.

### What causes winter damage?

Boxwoods are particularly vulnerable to injury during a winter following a dry summer or fall. Plants in poor health or those producing

*This winter damage, removed bark, is common for boxwood exposed to southern or western sun during winter months.*

foliage in the fall due to excessive rain, fertilizer or pruning are likely to be damaged because the plants do not have time to harden off. Winter damage is also common on branches that have developed aerial roots. Growing near the base of the branches, it will result in exposed roots that are subject to desiccation and freezing. When damage occurs the foliage will discolor. Removing the debris that accumulates in the center of the plant and keeping mulch away from plant stems will prevent the growth of aerial roots. Individual conditions are:

**Rapid temperature drop** can cause bark splitting in Zones 4 to 8. It is most likely to happen during the clear sunny days of January and February. The plant is warmed enough to start the production of new secondary vascular tissue immediately under the bark tissue. The plummeting temperatures at night cause the bark tissue to freeze and expand. This pushes out against the bark which splits and separates from the wood.

**Prolonged low temperatures** when the ground is deeply frozen will prevent roots from taking up water, making the boxwoods susceptible to desiccation. Once a hard freeze occurs, there are few alternatives. The impact can be reduced by applying mulch in late fall.

**High winds** will cause excessive transpiration which results in the plant losing unusually high amounts of moisture. An inadequate supply of soil moisture is often a contributing factor. Damage can be avoided by replenishing the moisture level.

**Snow and ice** can cause stems to bend, crack or even break. The snow ought to be removed from the plants as soon as practical. The snow is removed by gently shaking the shrub with a broom or stick in a side to side motion; do not use an up and down motion. If the temperature is under 35°F., the snow should not be removed. It is likely that the branches are frozen and will break. Ice should not be removed from the plant regardless of the air temperature.

### How can winter damage be avoided?

**Physical barriers,** while not aesthetically pleasing, are very useful. A snow fence frame placed over the top of the boxwood provides excellent protection. Burlap or lath fence will protect the plant by partially shading it from the sun and protecting it from the wind. These should be set up when the surface of the ground freezes and should be removed when temperature colder than 20° to 25°F is past. Physical barriers should not touch the boxwood foliage.

**Healthy, vigorous plants** are most able to withstand the stress of winter. This can be accomplished by maintaining adequate soil moisture, fall fertilization, and proper thinning. It is important to check that the center of the plant is free of dead

*Ice formation on* Buxus harlandii.

*Culture*

leaves and other debris. During dry periods throughout the year, water as necessary. Provide wind protection for plants in exposed sites.

**Mulch** applied in late fall will prevent rapid temperature changes at the soil surface and will prevent deep penetration of frost and excessive loss of surface water.

**String** will protect *B. sempervirens* 'Suffruticosa' from snow damage, by wrapping the outer branches. First tie the string securely to the main trunk at the base of the shrub. Then wrap the string in an upward spiral pressing the branches upwards and inwards. Work up to the top of the plant then back down and tie the string onto the trunk again.

The rows of string need to be about 8" to 10" apart to provide the best support. The tension applied to the string should be enough to prevent the branches from breaking under the weight of the snow, but not so tight that the air can not circulate through the plant.

**Antitranspirants** are chemicals that reduce transpiration, or water loss from the leaves.The most common type is an emulsion. Formulations may contain either wax, latex or plastics that form a film over the leaf which impedes water evaporation. Antitranspirants have limited effectiveness, lasting for 2 to 14 days. While beneficial in reducing water loss, the emulsions harm the plant by reducing the carbon dioxide that enters the leaf.

### How can winter injury be corrected?

Dead or broken stems should be removed by cutting them back to live wood. Branches that appear dead from winter injury may not be dead. To distinguish between a live and dead branch, use the thumb nail test. Take your thumb nail and scratch the bark off a small area in question. This will expose the vascular cambium tissue under the bark. If the tissue is green, the branch is alive, if brown, then it is dead and should be removed. On plants where the foliage has turned to a reddish-brown, delay pruning until after new growth starts in the spring.

On large boxwoods, the death or breakage of a large branch will leave an open area. Pulling limbs together to close the gap is not recommended. Ropes, twine and wire will crush the soft bark tissue, the tension will crack the branch crotches and it will discourage new foliage growth. Preferably, the gap should be left open to encourage new leaves which will naturally close the gap.

# Chapter 5

# Boxwood Propagation

*Young boxwoods, propagated from stem cuttings, growing in a greenhouse, planted in one-quart containers.*

# BOXWOOD PROPAGATION

There are occasions when additional boxwoods are desired for the garden. One way to obtain new plants is to purchase them. Another way is by propagation. This is a satisfying and quite enjoyable way to obtain new plants. The biggest advantage to propagation is the ability to obtain many plants from a single plant. There are several methods of propagation from which to choose. The choice should be dictated entirely by the desired results.

---

- Cuttings will produce plants that are identical to the parent plant. But the location on the plant from which the cuttings are taken will slightly influence the growth habit as the cutting develops into a plant. For example, if the cuttings are taken from the top of the plant the cutting will tend to grow more upright. If the cuttings are taken from a lower side branch the cutting will tend to have a spreading habit.
- Layering will produce a plant that is identical to the parent plant. This is the quickest method to produce a new plant.
- Seedlings are highly varied from the parent plant. This is advantageous when seeking out new and desirable characteristics in plants.

---

## CUTTINGS

Stem cuttings can be successfully taken from July to December. During this period, the cutting has a chance to harden off which will prevent wilting before a root system grows. Taken later in the winter, the cuttings are slower to root. The cutting procedure is quite easy.

Cuttings are taken from one year old branchlets. For example, if using *B. sempervirens* 'Suffruticosa' the cutting would be about $2^{1}/_{2}$"; if using *B. microphylla* var. *japonica* it would be about 4" long. Cuttings are best if obtained in the cool, early morning hours when the stems have the greatest concentration of water. Once collected, the leaves are removed from the bottom 1" of the cutting. This bottom portion can be treated with a rooting hormone. Nearly equal results are achieved in treatment or non-treatment with rooting hormones, which are intended to induce the cutting to develop a root system.

Nearly any type of container can be used as long as it is able to hold the media and provide drainage. There are several media mixes that are superior in promoting rapid and vigorous rooting. A good one is an

*Propagation*

equal portion by volume of pine bark; coarse, sharp builder's sand; and perlite. The cuttings are then ready to be put in flats or trays which are placed in shade.

Rooting usually occurs in two to three months. During this time, environments with high humidity consistently result in superior rooting. Frequent watering with a spray bottle to mist the cuttings provides satisfactory results. The plants can be planted out in a protected area the following spring.

## LAYERING

Layering occurs when roots develop on a stem while it is still attached to the parent plant. When it is pruned off, it becomes a new plant with its own root system. Some boxwoods will do this naturally, others can be easily induced.

### Natural layering

Look for low-growing branches. Other good candidates include branches that have fallen under their own weight or have been pulled down by snow. When in contact with the ground, these branches will grow their own fibrous roots. Growing from the ground near the dripline, the juvenile plant is a rapid grower, with a strong upright main leader. In contrast, the branches from the parent plant are more horizontal at the drip line. It is important to remove the layered growth whether or not this new plant will be used. If removed while the layer is only a

*A low-growing branch from a* B. microphylla *var.* japonica *has rooted into the soil.*

few years old, it can be cut off the parent plant, saved and transplanted. If permitted to grow, this vigorous growing plant will easily mis-shape the parent plant. In this situation neither plant will be able to grow properly due to the competition and stress of overcrowding.

**Induced layering**

First, select a one-year old branch or the tip of a branch near the ground. Put an object over the branch to keep it in contact with the soil.

*Once uncovered, the extensive root system is revealed. The crown of the plant is to the right. Note that the branch is larger in size as it grows to the left. This is a result of the additional water and nutrients absorbed by the adventitious roots.*

Make sure the branch is in contact with the soil, not the mulch. This is all that is necessary to induce layering. The preferred time to do this is spring. The branch will quickly develop its own root system. In the fall or next spring cut the branch between the crown and the layer. Now the new plant is ready to be transplanted.

## SEEDLINGS

Seeds are produced within the small fruit. The fruit capsule looks like an upside-down porridge-pot that is divided into three cells. There are two shiny black seeds in each cell. When the capsule dries, it splits open and the seeds fall out. Fruiting times vary from plant to plant and from year to year, so the seed collector must be on the alert for the opening and drying of the seed husks beginning in June.

Seeds will only be produced on a mature plant, one which is at least several years old. Some cultivars very rarely produce fruit, usually only under conditions of stress. A few cultivars of *B. sempervirens* and *B. microphylla* have never been observed in flower, while other cultivars that produce some flowers may produce no or only imperfect fruit.

An exciting aspect of growing boxwood from seeds is that the seedlings will not have all of the characteristics of the plant from which the seed was collected.

There is even a good method to grow seedlings without taking the time to collect, germinate and nurture young plants. Boxwood seedlings occur naturally under or near mature plants. These seedlings may be transplanted to individual pots or directly into a nursery area for further growth and observation. Care must be used in removing seedlings to avoid disturbing

*The mature seed capsule splits into three cells, exposing the six black seeds.*

*These seedlings, growing underneath a* B. microphylla *var.* japonica *are one year old.*

too much of the root system of the parent plant. Until the plants have become established in their new location, they should be protected from direct sunlight.

Boxwood seeds will germinate very easily and uniformly if given a cold stratification of 40°F for ten weeks. Seedlings that germinate in the spring can be planted out in protected areas the next spring. It will take about two or three years for a seedling to begin to exhibit its true characteristics such as leaf size and color and rate of growth. Plant habit characteristics such as shape and size will take much longer.

Boxwood flowers have a delicately sweet and pervasive fragrance, particularly on warm March and April days. This flower fragrance is much in contrast to the musky odor of the plant itself which is often debated by those who either love or despise it. Most boxwoods produce flowers in great profusion. From a distance the entire plant may take on a yellowish-green hue from the thousands of small blossoms borne in the leaf axils.

# Chapter 6
# Boxwood Pests

*While perhaps not immediately apparent, the lack of leaves on this* Buxus microphylla *var.* japonica *'National' is due to* Monarthropalpus flavus, *the Boxwood Leafminer. The leafminer damage was high enough to cause the plant to drop a full year's growth of leaves.*

# BOXWOOD PESTS

Insects belong to a major group of the animal kingdom named the Arthropods. Gardeners commonly encounter insects, which are in the class Insecta, and mites, which belong to the class Arachnida. Insects are the most abundant and most successful form of life on the earth. There are about one million species of insects that have been identified. But fortunately, only about 10% of these can be considered pests, and only five have a serious impact on boxwood.

Competition between people and pests has increased with the increase in human population. As a result, the management of pests has become more important and focuses on some broad questions.

### Are control methods needed at this time?

It is important not to begin control methods unless pests are actually present, at a vulnerable stage in their lifecycle, and causing an intolerable amount of damage. This can be properly answered if there is regular monitoring of the plants for damage in the presence of an insect or mite pest. It involves knowing the life cycle of the pest so that control measures can be taken when it is most effective. An example is the cupping of boxwood leaves which occurs in June. Applying an insecticide at that time will provide no control because the psyllid has finished feeding and there will be no additional damage that year.

### What is a scouting map?

A map showing the location of boxwoods and other plants in the garden is of tremendous value. It can be used to track pest infestations and evaluate the effects of control actions. Examine plants carefully every one to two weeks in the growing season to see if any insects or mites are present on the plants. Mark the location of infested plants on the map and note the severity or extent of the pest problem. It is important to know that the majority of the time, pests favor certain parts of the plant or plants in one location over another.

So, if a scouting map shows the psyllid population is increasing in one portion of the garden, it would be in order to spray only the infested plants. Treating the entire garden would not be appropriate. In this way beneficial insects such as lady beetles are not eliminated from the entire garden.

### How can I best control this pest?

Before a pest can be reliably controlled it should be properly identified. If the plant damage level becomes objectionable, action is in order. Management tactics available for control use one or a combination of strategies.

### Mechanical control

Prune off the infested leaves or twigs. This is practical for small localized infestations.

### Cultural control

Creating a scouting map will permit you to look at the boxwoods in a way you never have before. This leads to determining the causes of the damage. If plants are properly maintained, the damage produced by insect or disease infestations may be reduced or better tolerated. The best approach involves a long-term strategy that reduces the cultural stress that can affect boxwoods.

### Chemical control

Horticultural spray oil can be used in the dormant season to reduce overwintering populations of mites and armored scales. This is a contact material and coverage must be thorough to obtain good results. It is environmentally safe to use and is inactive once the material is dry. It may also be used at reduced rates in the growing season for most pests. Insecticidal soaps are effective for most pests when used during the growing season. The soaps, like the oil, must be thoroughly sprayed to ensure contact and are inactivated once they dry on the plant. Best results are obtained when the mites, scale crawlers and other pests are actively feeding in the open and are accessible to contact with the soap.

In the application of spray materials, it is important to understand that the boxwood leaves are hydrophobic. That is, the thick waxy leaves of boxwood will repel water. Insecticides are hydrophilic, and do not

---

The pests are listed below in order of their ability to damage boxwood. For example, mites can be expected to be a more serious problem than scale insects.

| | |
|---|---|
| **Leafminer** - Blistering of young leaves in late summer | (see p. 64) |
| **Mites** - Stippling of older leaves in summer | (see p. 66) |
| **Psyllid** - Cupping of the new leaves in spring | (see p. 68) |
| **Scale** - White or brown, waxy bumps on twigs in late summer to early spring | (see p. 69) |
| **Webworm** - Webs on the inner branches in summer | (see p. 70) |
| **Wildlife** - Variable, depending on the animal | (see p. 71) |

easily adhere to the leaf surface. The addition of a surfactant to the insecticide greatly improves the ability of the spray material to remain on the leaf. This provides much more effective results when applying insecticides.

## LEAFMINER

The boxwood leafminer, *Monarthropalpus flavus*, is actually a gall midge and not a leafminer. It is a serious insect pest of boxwood. A high population can defoliate and kill boxwoods. Injury is caused by the larvae (maggots) feeding in the leaf and resulting in premature leaf drop. Most cultivars of *B. sempervirens* and *B. microphylla* are susceptible to the leaf miner.

*Leafminer larvae damage on leaf.*

**What does leafminer damage look like?**

The boxwood leafminer feeds on the tissue between the top and bottom of the leaf. The resulting damage appears as an irregular oval swelling on the leaf. There may be a slight blistering of the leaf on the lower surface with a yellowish or brown discoloration. The new foliage will not show this blistering effect until late summer. The female, a true bug, enters the leaf through the stoma, a natural opening in the leaf surface used for the exchange of air. Early signs are holes on the lower leaf surface after the female deposits her eggs. These punctures are visible if the leaves are closely inspected. Leafminers prefer the protected part of the boxwood, the lower and innermost new leaves. Damage from high infestations results in premature leaf-drop.

### Understanding the leafminer life cycle

The leafminer produces a single generation each year. They overwinter as translucent yellowish-green, partly grown larvae inside the leaves. During the early, warm days of spring they grow rapidly into

*Pupal casing left by emerging adult.*

yellow-orange colored pupae. They time their pupation so that the adults emerge from the underside of the leaf as flies in late April usually when *Weigela* begins to bloom. The adult fly is a little shorter than ¹/₂" long and re-sembles an orange mosquito. The flies emerge over a three-

*With the entire lower leaf surface removed, the developing leafminer larvae can be seen.*

week period, leaving noticeable white pupal skins which hang down from the undersurface of the leaf.

It is easy to notice the yellow-to-orange colored adults flying

*Pests*

around from leaf to leaf, particularly if the plant is disturbed. The females quickly begin laying small white eggs deep into the leaf tissue from the under side of the new spring leaves. The entry points, tiny egg-laying ruptures, can be seen on the underside of the leaf. One female will deposit about 30 eggs, often dying within hours of laying her last eggs. The eggs will develop for three to four weeks. The resulting larvae will feed inside the leaf all summer, fall, and early the following spring.

During feeding, the larvae use a microscopic hook to rupture the leaf cells. The plant responds by sending more biomass into the leaf, which gives the larvae more to eat.

### How to control leafminer

Control is necessary when persistent or intolerable damage is observed. The leafminer is most susceptible right after the adult emerges and before the eggs are laid. The adults emerge over a three-week period, but each lives for only two days.

Control measures should be scheduled when adults are seen, usually late April to early May. Contact insecticides are effective when used against the adult. Systemic insecticides may be used against the first instar of the larvae in mid-June.

Birds, apparently able to hear the larvae inside the leaf, have been known to attack high populations of larvae. While cold temperature does not cause mortality to the over-wintering larvae, they will dry up and die through the lack of moisture during hot, dry summers.

## MITES

The mite is not an insect, but in a separate class, the *Arachnida*. The boxwood mite, *Eurytetranychus buxi*, is a rather common and wide-spread pest. Mites are very inconspicuous. Early damage is also not obvious and the problem is often overlooked until high populations and damage have been reached.

The mite is a serious problem on most *B. sempervirens* cultivars, particularly those grown in sunny locations. It also attacks *B. microphylla* var. *japonica*.

### What does mite damage look like?

The mite feeds on the upper and lower leaf surfaces. Depending on how large the population is, it will produce varying degrees of leaf stippling. On close examination, the damaged leaf appears as if someone took a small needle and made tiny white scratch marks. This damage is a result of the mites piercing the epidermal cells of the underside of the foliage, sucking out the juices of the new leaves. In the absence of the

*Damage by the boxwood mite is most noticeable on older leaves.*

green chlorophyll pigments the surface of the leaf appears dusty or pale green. The mite damage is most apparent on the second and third year leaves. Viewed from a short distance, the damaged plant takes on a light silvery cast.

### Understanding the mite life cycle

The mite overwinters as a yellowish green egg on the underside of the boxwood leaf. The eggs hatch in April and May. The adult is period-sized, yellowish-brown, has eight long legs and closely resembles a miniature spider. Mites favor hot, dry conditions. For example, at 70°F, mites complete their life cycle, egg-to-adult, in as little as ten days. The population increase is greatest in early summer. During a hot and dry summer, it is common to have five to eight generations of mites. This can result in a rapid buildup of populations and extensive damage.

### How to control mites

Many species of insects and other mites feed on the boxwood mites. The presence of mites can be determined by lightly "beating" a small branch over a white sheet of paper. The mites will fall onto the card and can be seen moving around. It is important to delay control measures until mites reach a damaging level. Damage occurs at about 25 mites per "beat." This "25-mite" level permits predatory insects and mites to become established. Horticultural oil is effective if applied at a summer rate. It will kill eggs and adult mites.

## PSYLLID

The most common insect which attacks *B. sempervirens* and its cultivars is the box-wood psyllid, *Psylla buxi*.

### What does psyllid damage look like?

The nymphal insect feeds in the spring when boxwoods produce new leaves. It consumes leaf cells, causing the young leaves to become distorted and cupped inward. The damage is quite conspicu-

*Spring growth damaged by psyllid.*

ous. This condition is unsightly and will stop growth at the shoot apex for a 2- to 3-year period. Large populations can kill the young leaves, but do not otherwise damage the plant.

### Understanding the psyllid life cycle

There is one generation of psyllids per year, they overwinter as

*White secretion from the psyllid deposited at new growth.*

*The adult psyllid is small and nearly the same color as the leaf.*

orange eggs beneath the bud scales. The $^1/_8$"-long light green nymph emerges in mid to late April, coinciding with the emergence of the spring foliage. The nymphs produce a white, waxy, string-like secretion while they feed within the cupped leaves, until the adults appear in June. The adult does not feed and simply flies around depositing eggs into the bud scales during June and July.

*Pests*

### How to control psyllid

Low populations are common and injury is usually not serious enough for control measures. Control is used to reduce the population and their damage to a level that will not affect the growth and overall health of the boxwood. Correct timing is most effective in reducing populations. Good control can be obtained by using insecticidal soaps. Make an initial application as the new growth emerges in April, with a second application three weeks later.

## SCALE

There are a number of different scale insects that attack boxwoods. They include: the wax scale which has common names that include Japanese, Indian and ceriferus scale (*Ceroplastes ceriferus*), oystershell scale (*Lepidosaphes ulmi*), California red scale (*Aonidiella aurantii*), Oleander scale (*Aspidiotus*

*The large, white covering of the female wax scale is found on the branches inside the foliage.*

nerii), peony scale (*Pseudaonidia paeonia*), greedy scale (*Hemiberlesia rapax*), euonymus scale (*Unaspis euonymi*), San Jose scale (*Quadraspidiotus perniciosus*), cottony maple scale (*Pulvinaria innumerabilis*), and the cottony cushion scale (*Icerya purchasi*). The majority of these scales occur only in the southern region of the United States, and are of particular concern in California and the states bordering on the Gulf of Mexico.

For the majority of boxwoods grown in more temperate northerly regions, these scales are only occasional problems and are not generally destructive. The one exception is the wax scale which attacks boxwood almost everywhere.

While most common in the southern United States, it is found along the East Coast to a northern limit of New York. As a result of this large distribution and its potential for damage, attention here will focus on the wax scale.

## Understanding the scale life cycle

The wax scale has only one generation per year. In the deep south, two generations per year are possible. It overwinters as an adult female which can be expected to lay about 1,000 eggs. The mature female is about $1/3$" long and has a soft, bright white, waxy covering. The eggs are laid in April or May and hatch in June.

The emerging crawlers will settle on the twigs and stems, but not on the leaves. Soon after they molt the nymphs begin producing wax which creates a cameo appearance. As wax production continues the body is quickly and fully covered with wax by the end of July and chemical control is impractical.

## How to control scale

Correct timing of treatments will provide superior results. The eggs hatch in early June and the emerging crawlers are quite active. This is the best time to apply controls that should cover all of the trunk, stems and branches. Foliar sprays of horticultural oil, insecticidal soap, or residual insecticides may be used.

# WEBWORM

Compared to all of the other pests, the boxwood webworm is a minor problem. However, it does have regional population outbreaks that make this insect important enough to mention here. The primary host of the webworm is *B. sempervirens* 'Suffruticosa' and other dense and compact plants such as the slow-growing cultivars of *B. microphylla*.

## What does webworm damage look like?

The larvae spin loose webs along the stems and twigs while feeding on the innermost leaves. Since many spiders inhabit the interior portions of boxwoods, the webbing may be a bit confusing. The real question is whether these are spider webs or webworm webs. The webworm will have fecal pellets and partially chewed brown leaves scattered through the webbing; spider webs will not.

## Understanding the webworm life cycle

A shy and inconspicuous insect, the webworm prefers to stay on the inner parts of very dense plants. It has only one generation each year. The webworm overwinters as a grayish larva that will reach up to $1/2''$ long before becoming an adult in May and June. The adult lays eggs during May and June. The small larvae feed during July and August.

### How to control webworm

Usually, measures to control the webworm are not necessary because population levels are seldom high enough to necessitate any action. Manually remove and destroy webs and caterpillars in small infestations.

## WILDLIFE

While not often considered to be pests, the damaging effect of some wildlife can be costly and discouraging for any gardener.

### Dogs

Dogs can damage boxwoods by digging up surface roots which are easily injured or killed. Dog urine on boxwoods creates a circular zone of dead foliage near the base of the plant. The urine will kill the foliage but does not damage the twigs. New foliage will quickly grow after this type of damage. While repeated occurrences of digging or urination can cause partial branch dieback, controlling domesticated dogs is not practical.

### Voles

A small mouse-like rodent, there are two common types: the meadow vole, known as the meadow or field mouse, and the pine vole, called the pine mouse. Their populations may vary greatly from year to year. The dense boxwood foliage close to the ground provides an ideal habitat. Voles eat the root and bark, stripping it bare from the trunk and branches near the ground. Most active on boxwood during fall and winter, they can girdle and kill a plant. Control is effective using several techniques: keep the mulch depth at one inch or less; mow the grass around the plants. A young, heathy cat that is well fed and secure in its territory is an excellent predator.

### Ponies

Occasional grazing results in temporary stunting, and constant feeding can remove enough foliage to kill a boxwood. Boxwood contains the poisonous alkaloid, buxene. There have been numerous cases of young horses eating large quantities of boxwood, without suffering any ill effects. While ponies are apparently not at risk by eating boxwood, each should be separated from the other by a barrier such as a fence.

### Deer

The white-tailed deer can be found through most of the United States. Residential and other building sites are occurring in more rural areas that were once habitat areas for deer. This has resulted in a need for ornamental plants that will not be injured by deer. Fortunately, boxwood are deer-resistant. Deer will not eat boxwood unless very hungry and no other food source can be found. However, deer will occasionally damage boxwood by rubbing their antlers on the bark. Since they do not usually damage boxwood severely, controlling deer, or protecting the boxwood, is not necessary.

# Chapter 7
# Boxwood Diseases

*A serious disease affecting English Boxwood is* Volutella. *A canker disease, it commonly attacks branch and stem tissue. It results in the death of the entire system.*

Fungi are responsible for more plant diseases than any other group of organisms and cause most of the diseases in boxwood. Fungi are small, generally microscopic in size, with a plant-like body. Since fungi lack chlorophyll, they spend part of their lives taking food from host plants.

Keeping boxwood healthy greatly reduces the potential for disease. By properly selecting a boxwood cultivar, which is then placed in the appropriate site and properly cared for, the impact of disease will be reduced. The important aspect of managing diseases is early recognition and control. Frequently, the early signs and symptoms of a disease are overlooked. Once the disease has been diagnosed, prompt corrective measures will give the most effective results.

The diseases of boxwood can be grouped into two categories; those that affect the foliage or twig and those that affect the roots. Both produce symptoms that appear on the foliage. Listed below are the diseases that have a major effect on boxwood and their symptoms:

---

*Phytophthora* - wilting and discoloring
   of the foliage                                         (see p. 74)
*Macrophoma* **leaf spot** - tiny, black,
   raised spots on the under-side of the leaf   (see p. 75)
*Volutella,* **or canker** - a browning of
   leaves and twigs, pink fuzz-like growth
   on the underside of the leaf, or a
   wound sunken beneath the bark               (see p. 77)
**Nematodes** - wilting, stunting, yellowing
   of the foliage                                         (see p. 79)
**Decline** - plant growing poorly, small
   leaves (some defoliation or dieback
   may be present)                                       (see p. 80)

---

## *PHYTOPHTHORA*

Phytophthora means plant destroyer, which gives an indication of how serious a problem this can be. There are about 40 species, but only one is important to boxwood, *Phytophthora parasitica*. This soil-borne fungus affects all cultivars of *B. sempervirens* at any age or size; it damages leaves, stems and roots. Infection begins when the soil is wet and cool, about 58° to 70°F. This type of weather is common in the spring and fall. The disease will progress and cause injury only under higher soil temperatures. Damage occurs at 75°F, with the greatest effects occurring at 85°F. Boxwoods infected with *Phytophthora* seldom survive. *Phytophthora* produces spores that move in a water film in the soil. As a result, *Phytophthora* is most damaging to boxwoods growing in poorly drained soils.

## What does it look like?

The leaves will first gradually turn yellow, then the edges will become wavy. Next the leaves will change to a bright straw color and remain attached to the twigs. This may happen to one branch, or several, or throughout the entire shrub simultaneously.

The stem will have a dark brownish-black coloration of the vascular tissues under the bark at ground level to a few inches above. The dark color is a response of the vascular tissue to the fungus. The fungus can cause a partial, or even complete, blockage of nutrient and water movement in the stem.

The roots will appear dull and dark brown in color. This is in contrast to healthy roots that will have a bright, light tan color. The "bark" will be decayed and is easy to remove.

## How can I correct the problem?

### Avoid the initial infection
Before the boxwood is planted, first evaluate the site. Plant boxwoods in soil that has adequate internal drainage. Wet soil conditions should be avoided, particularly in the spring and fall.

### Boxwood weakened by *Phytophthora*
Transplant the boxwood to a site that is well drained. If transplanting is not practical, consider drains and/or grade changes to redirect water away from the boxwood.

### Boxwood killed by *Phytophthora*
The boxwood should be removed from the site. The area should not be replanted until soil drainage is improved.

## *MACROPHOMA* LEAF SPOT

This imperfect fungus, *Macrophoma candolleri*, attacks weakened or decaying branches of many cultivars of *B. sempervirens*, especially 'Suffruticosa'. Fortunately *Macrophoma* is only of minor concern and is classified as a secondary invader or a weak parasite. Usually the fungus infects plants that have been weakened by poor culture, winter-burn or have overly-thick foliage.

## What does this leaf spot look like?

On close inspection tiny, black, raised spots can be found on the under side of either the light green or, more commonly, tan-colored foliage. These are the round and hollow fruiting bodies, or pycnidia, of the fungi. The large, transparent, single-celled conidia are produced in large quantities and ooze out of the pycnidium when placed in water. Water is the primary means of dispersal and movement for fungi. The fruiting bodies first appear on the oldest leaves inside the center of

*A characteristic sign of macrophoma leaf spot is the tiny, black fruiting bodies.*

the plant. As the infestation progresses it affects the younger leaves which will defoliate. Heavy infestations can cause entire branches to die in only a few weeks.

## What can I do about *Macrophoma*?

Typically the *Macrophoma* leaf spot can be corrected by pruning out the infected branches when the pycnidia appear. In addition to pruning out the fungus, the removal of branches will improve the air circulation through the plant which will help contain and eliminate the fungus. The fungus prefers a moist, cool, dark area which can be found in the center of dense boxwood plants.

The real secret of control is not to let the *Macrophoma* leaf spot get started. Preventive measures are always far easier to perform than trying to heal an infected boxwood.

Thinning boxwood will permit both air and sunlight to circulate freely through the plant, providing an unfavorable site for the fungus to establish itself. Boxwood growing in a sunny location are more likely to develop overly-thick foliage. Sanitation is an additional preventive measure. Removing infected foliage will prevent reinfestation of the *Macrophoma*.

## VOLUTELLA

This fungus, *Volutella buxi*, is stem blight or canker. It primarily affects *B. sempervirens*. It is the most serious disease to affect mature and overly thick *B. sempervirens* 'Suffruticosa'. It is especially severe in periods of high humidity. A heavy infestation results in defoliation and the death of entire stems.

### What does it look like?

In moist weather, a mass of creamy or light pink dust or fuzz-like growth is visible on the underside of the leaf. These colorful spores are very distinctive. This stage is often overlooked because it forms in the most dense portions of the plant. The pink spore masses may not be visible under dry conditions.

As the infection continues, the outer green leaves will become quickly and progressively discolored, changing to a dark brown and then tan color. The branch appears and smells almost as if diesel fuel had been poured on it. As the disease progresses, the entire branch will drop all of its leaves.

*Spores at actual size on the underside of the leaf.*

*Spores at 25x actual size.*

*Diseases*

The *Volutella* will usually cause the soft tissue of the current year's growth to discolor black. If the plant is weak, the discoloration will extend well into the previous year's growth, producing a stem canker.

A canker is formed from a wound or a dead, discolored area that often sinks beneath the bark on the stem. On the trunk and large branches, the healthy tissue immediately next to the canker may slightly increase in thickness and appear higher than the normal surface. Cankers will produce a crack on the surface of the bark that may be several inches in length. These cracks cause wilting and death to the parts of the branch beyond the canker.

### How can I correct this problem?

First, be sure that the problem is *Volutella buxi*. If the pink masses of conidia are not visible, they can easily be cultured. To do this, prune a small twig or take

Volutella *in an advanced stage on* Buxus sempervirens.

several leaves from a portion of the plant that appears to have *Volutella* symptoms. Wet the sample with ordinary water. Place it in a zip-lock plastic bag and keep the sample in a cool area. After a week or two, fuzzy pink colonies will develop if it is *Volutella*.

### Avoid the initial infection.

Do not water boxwood with oscillators or other methods that wet the foliage. Frequent rain or overhead watering encourages establishment of *Volutella*. Maintain a properly thinned boxwood to improve air circulation, which quickly dries the plant. Keep the boxwood in vigorous health by following good cultural practices.

### Control established infections.

Prune the diseased branch 6" to 12" below the affected tissue. This is best done during dry weather to prevent spreading the pycnidia. Removing the diseased branch from the site will provide nearly total control.

# NEMATODES

Nematodes are small eel-like worms that cause plant disease. Nematodes live as an obligate parasite which can only grow and multiply when on a living host. However, they can live as an egg or cyst for several years without the benefit of a host.

There are thousands of nematode species, but only two are important to boxwood. Both commonly occur on boxwood in Maryland, Virginia and North Carolina.

## How do I know if the boxwoods have nematodes?

Nematodes feed on the roots of boxwood. The first visible symptoms are wilting, stunting, yellowing and/or bronzing of the foliage. Symptoms appear fairly uniformly throughout a plant, not branch-by-branch as in *Volutella*.

If boxwoods are growing in proper cultural conditions and still exhibit these symptoms then nematodes are suspect. The only reliable method of determining the presence of nematodes is to collect soil and root samples from a suspect boxwood and have the sample examined by a plant nematologist.

## Which are important to boxwood?

Lesion or Meadow nematodes, *Pratylenchus spp.*, are very destructive on *B. sempervirens* cultivars and *B. microphylla* var. *japonica*. These nematodes prefer to feed on the smallest roots; this has a root-pruning effect, which may be followed by root rotting from a secondary infestation of pathogens. Above-ground damage is slow to appear, but results in smaller leaves that are yellow or, more often, reddish. The branches may keep only one or two years of foliage.

Root knot nematodes, *Meloidogyne spp.*, cause swelling of the roots. The formation of a gall stops root growth, which reduces the availability of nutrients and water. The plant will be stunted with reduced shoot growth, small yellowish or bronze-colored leaves.

## How are nematodes controlled?

Nematodes cannot be totally eliminated from the landscape. Solarization of soil prior to planting is effective. The goal is to keep the population low enough to prevent damaging symptoms that weaken the plant. Boxwood should not be grown in soils heavily infested with nematodes. In the long term, growing plants such as grasses that are not affected by nematodes will reduce nematode populations.

Current control options for the lesion nematodes are either not effective or practical in the landscape. A biological control, *Bacillus pentrans*, has been effective when treating the soil for root knot nematodes.

## DECLINE

While the pathogen species is a topic of much debate, *Paecilomyces buxi* (formerly *Verticillium buxi*) is presumed responsible for boxwood decline. Decline is thought to be the result of fungi and/or nematodes that invade the root and/or crown portions of the boxwood that are culturally weak. There does appear to be a complex of several fungi, parasitic nematodes, environmental and cultural factors associated with decline. Their specific interaction with each other and their association with the decline is not clearly understood. Decline is limited to *B. sempervirens* 'Suffruticosa'.

The distribution of decline extends from New York to North Carolina and the Appalachian Mountains. While numerous cases of decline were reported in the 1970s, there have been few since.

### What does decline look like?

The deepest roots are affected first. As the disease progresses, the stem below the ground begins to turn brown. This will extend upward into branches, often in a random pattern. The above-ground symptoms take the form of small leaves that are brittle and yellow or red-colored as well as characteristic browning of one branch or several branches in a random fashion. In severe cases, there is a sudden wilting of foliage or dieback of entire branches, resulting in the death of the boxwood.

### How can I correct the problem?

Boxwoods in poor health are susceptible to decline. There is no tested, proven treatment for boxwood infected with decline. Initial infections can be avoided by keeping *B. sempervirens* 'Suffruticosa' culturally healthy.

# Appendix A
# Species, Cultivars and Hybrids

Boxwood is in the botanical family known as *Buxaceae*. The other genera included in this family are: *Pachysandra*, *Sarcococca* and *Styloceras*. *Simmondsia* had been classified in this family, but was reassigned to its own family in 1984.

There are about 90 species of boxwood, most of which grow in the tropical regions of the world. The natural distribution of boxwood includes three large areas. The first is southern Europe, Germany, Spain, Switzerland, North Africa extending to Iran and the Caucasus. The second area embraces the islands of the West Indies. The third area, East Asia, includes the Himalayas, China, Korea and Japan.

Following is a list of the cultivars and hybrids of *Buxus*. The cultivar name is followed by the first valid citation found in literature. Invalid cultivar names include their valid cultivar synonyms. Valid cultivar names appear in **boldface type** while invalid and unregistered names are in lightface type. Botanical names appear in *italic type*.

---

***Buxus balearica*** Lam. in *Encyc. Meth. Bot.* 1:511.1785.
*'Marginata'* P. Corbelli in *Dizionario di Floricultura* 1:231.1873.

---

***Buxus bodinieri*** Lev. in *Fedde Repertorium* 11:549.1913
'David's Gold' Stone House Cottage Nurseries, Worcestershire, England 1991

---

***Buxus harlandii*** Hance, Supplement to the Flora Hongkongensis in *Journal of Linnean Society* 13:123-124.1873.
**'Richard'** J. Baldwin in *The Boxwood Bulletin* 2(4):44.1963.

---

***Buxus microphylla*** Sieb. & Zucc. in *Abhandl. Math. Phys. Konigl. Akad. Wissench. Munch.* 4(2):142.1845.
'Asiatic Winter Gem' Listed in numerous nursery catalogs = **'Winter Gem'**
**'Compacta'** D. Wyman in *American Nurseryman* 107(7):50.1963.
'Creepy' Oliver Nurseries, 1159 Bronson Road, Fairfield, CT 1986.
**'Curly Locks'** D. Wyman in *American Nurseryman* 107(7):50.1963.
'Fiorii' Fiore Enterprises, Rt. 22, Praire View, IL 60069 ?
**'Grace Hendrick Phillips'** H. Hohman in *The Boxwood Bulletin* 7(1):1.1967.
**'Green Pillow'** O'Connor in *Baileya* 1:114.1963.

'Green Sofa' J. Baldwin in *The Boxwood Bulletin* 15(3):42.1976.

'Helen Whiting' J. Baldwin in *The Boxwood Bulletin* 15(3):41-42.1976.

'Henry Hohman' J. Baldwin, College of William & Mary, Williamsburg, VA. ?

'Jim's Spreader' Garrison's Nursery, Seabreeze, NJ. ?

'John Baldwin' P. Larson in *The Boxwood Bulletin* 28(2):27.1988.

'Kingsville Dwarf' D. Wyman in *American Nurseryman* 107(7):50.1963 = 'Compacta'

'Locket' J. Baldwin in *The Boxwood Bulletin* 15(3):41.1976 & 16(1):10-11.1976.

'Miss Jones' R. Jones, Rt. 2, Box 93, Eatonton, GA. 1967.

'Sunlight' M. Gamble in *The Boxwood Bulletin* 28(2):26.1988.

'Sunnyside' Sunnyside Nurseries, Troy IL.?

'Winter Gem' John Vermeulen & Son, Box 267, Neshanic Station, N.J.1982.

*B. microphylla* Sieb. & Zucc. var. *asiatic* 'Winter Gem' Listed in numerous nursery catalogs = B. *microphylla* Sieb. & Zucc. **'Winter Gem'**

*B. microphylla* Sieb. & Zucc. var. *koreana* Nakai ex Rehder in *Journal of the Arnold Arboretum* 1:35.1919. nom. -Rehder in *Journal of the Arnold Arboretum* 7:240.1926. = **B. sinica var. insularis**

*B. microphylla* Sieb. & Zucc. var. *insularis* Nakai in *Botanical Magazine of* Tokyo 36:63.1922. = **B. sinica var. insularis**

---

**Buxus microphylla** Sieb. & Zucc. var. **japonica** (Muell.) Rehd. & Wils. in Sargent, *Plantae Wilsonianae* 2(1):168.1914.

'Alba' Catalog, Andorra Nurseries, Chestnut Hill, Philadelphia, PA. 1908.

'Angustifolia' L.H. Bailey in *Hortus* 105.1930.

'Argentea' Beissner, Schelle and Zabel in *Handbuch der Laubholz-Benennung* 283.1903.

'Aurea' Catalog, Charles Dietriche, Angers, France.1892.

'Fortunei' Catalog, Andorra Nurseries, Chestnut Hill, Philadelphia, PA.1908.

'Green Beauty' Sheridan Nurseries, Georgetown, Canada. Selected in 1957.

'Green Jade' Langley Boxwood Nursery, Hampshire, England.1991

'Japanese Globe' Plant List, Kelly Howell, Spokane, WA.1958.

'Latifolia' Catalog, Andorra Nurseries, Chestnut Hill, Philadelphia, PA.1908.

'Morris Dwarf' B. Wagenknecht in *The Boxwood Bulletin* 11(3):45.1972.

**'Morris Midget'** B. Wagenknecht in *The Boxwood Bulletin*
11(3):45.1972.
**'Nana'** Beissner, Schelle and Zabel in *Handbuch der Laubholz-
Benennung* 283.1903.
**'Nana Compacta'** Catalog, Mayfair Nurseries, Bergenfield, N.J.1954.
**'National'** D. Anberg in *The Boxwood Bulletin* 12(4):62.1973.
**'Obcordata'** Beissner, Schelle and Zabel in *Handbuch der Laubholz-
Benennung* 283.1903.
**'Obcordata Variegata'** Anonymous in "Lists of Plants Introduced by
Robert Fortune from Japan." *Gardeners Chronicle* 735.1861.
**'Rotundifolia'** Beissner, Schelle and Zabel in *Handbuch der Laubholz-
Benennung* 283.1903.
**'Rotundifolia Glauca'** Catalog, Chas. Dietrich, Angers, France.1892.
**'Rotundifolia Pendula'** Catalog, Andorra Nurseries, Chestnut Hill,
Philadelphia, PA.1919.
**'Rubra'** T. Makino in *Botanical Magazine of Tokyo* 27:112.1913.
**'Variegata'** L. Dippel in *Handbuch der Laubholzkunde* 3:83.1893.

---

*Buxus natalensis* (Oliv.)Hutchinson *Gen.Fl.Pl.* 2:108.1967.
'New Silver' Daisy Hill Nurseries Ltd., Down, N. Ireland.1991

---

*Buxus sempervirens* L. *Species Plantarum* 983.1753.
**'Abilene'** Inventory, Beal-Garfield Botanic Garden, East Lansing,
MI.1960.
'Acuminata' *Journal of the Royal Horticultural Society* 18:86.1895 = *B.
acuminata* J.Muller, Arg. Buxaceae. *De Candolle Prodromus*
16(1):15.1869.
**'Agram'** Introduced by the USDA, Glenn Dale Plant Introduction
Station, Glenn Dale, MD.1959.
'Albo-marginata' D. Wyman in *American Nurseryman* 117(7):57.1963
= **'Marginata'**
'Anderson' Edgar Anderson, St. Louis, MO.
'Andersoni' A name applied to a group of seedlings by Anderson. No
precise application of name seems possible.
**'Angustifolia'** P. Miller, *Gardener's Dictionary* ed.8:Bux.no.2.1756.
'Angustifolia Variegata' J. Loudon, *Arboretum et Fruticum Britannicum*
III:1333.1838. = **'Marginata'**
'Angustifolia Variegata Maculata' H. Baillon, *Monographie des
Buxacées et des Stylocérées* 61.1859. = **'Marginata'**
**'Angustifolia Variegata Punctulata'** H. Baillon, *Monographie des
Buxacées et des Stylocérées* 61.1859.
**'Arborescens'** P. Miller, *Gardener's Dictionary* ed. 8:Bux.no.1.1756.

'**Arborescens Argentea**' J. Loudon, *Arboretum et Fruticum Britannicum* III:1333.1838.

'Arborescens Aurea' J.Loudon, *Arboretum et Fruticum Britannicum* III:1333.1838. = '**Aureo-variegata**'

'**Arborescens Aurea Acuminata**' H. Baillon, *Monographie des Buxacées et des Stylocérées* 60.1859.

'Arborescens Aurea Maculata' H. Baillon, *Monographie des Buxacées et des Stylocérées* 60.1859. = '**Aureo-variegata**'

'Arborescens Aurea Marginata' H. Baillon, *Monographie des Buxacées et des Stylocérées* 61.1859. = '**Marginata**'

'**Arborescens Aurea Punctulata**' H. Baillon, *Monographie des Buxacées et des Stylocérées* 60.1859.

'**Arborescens Decussata**' *Kew Handlist of Trees and Shrubs* 269.1925.

'Arborescens Gable' Catalog, Tingle Nurseries, Pittsville, MD.1963. = '**Joe Gable**'

'Arborescens Longifolia' L. Dippel, *Handbuch der Laubholzkunde* 3:82.1893. = '**Angustifolia**'

'Arborescens Marginata' J. Loudon, *Arboretum et Fruticum Britannicum* III:1333.1838. = '**Marginata**'

'Arborescens Salicifolia' L. Dippel, *Handbuch der Laubholzkunde* 3:82.1893. = '**Salicifolia**'

'Arborescens Tenuifolia' L. Dippel, *Handbuch der Laubholzkunde* 3:82.1893. = '**Angustifolia**'

'Arborescens Thymifolia' H. Vogel, *Gartenwelt* 33:150.1929. = '**Thymifolia**'

'Arborescens Variegata' Catalog, Andorra Nurseries, Chestnut Hill. Phil., PA.1908 = '**Argenteo-variegata**'

'Argentea' C. Ludwig, *Die Neuere Wilde Baumzucht* 9.1783. = '**Argenteo-variegata**'

'Argenteo-marginata' L. Dippel, *Handbuch der Laubholzkunde* 3:81.1893. = '**Argenteo-variegata**'

'**Argentea Nova**' Catalog, V.Gauntlett, Chiddingfold, Surrey, England.1930.

'**Argenteo-variegata**' R. Weston, *Botanicus Universalis* 1:31.1770.

'**Aristocrat**' J. Baldwin in *The Boxwood Bulletin* 6(2):23.1966.

'Asheville' Inventory, Secrest Arboretum, Wooster, OH.

'Aurea' J. Loudon, *Arboretum et Fruticum Britannicum* III:1333.1838. = '**Aureo**-variegata'

'Aurea Maculata' *Kew Handlist of Trees and Shrubs* 131.1896. = '**Aureo-variegata**'

'Aurea Maculata Aurea' Inventory, Beal-Garfield Botanic Garden, East Lansing, MI.1960. = '**Aureo-variegata**'

'Aurea Maculata Pendula' Inventory, Beal-Garfield Botanic Garden, East Lansing, MI.1960. = **'Aurea Pendula'**

'Aurea Marginata' J. Loudon, *Encyclopedia of the Trees and Shrubs of Great Britain* 703.1853. = **'Marginata'**

**'Aurea Pendula'** *Kew Handlist of Trees and Shrubs* 131.1896.

'Aureo-limbata' R. Weston, *Botanicus Universalis* 1:31.1770. = **'Marginata'**

'Aureo-maculata' Dallimore *Holly, Yew & Box*.1908.

'Aureo-marginata' Beissner, Schelle, & Zabel, *Handbuch der Laubholz-Benennung* 284.1903. = **'Marginata'**

**'Aureo-variegata'** R. Weston, *Botanicus Universalis* 1:31.1770.

'Balkan' H.E. Nursery, 1302 East Union, Litchfield, IL 1983.

'Bass' M. Dirr *Manual of Woody Landscape Plants*.1983.

**'Belleville'** R. Siebert in *Arnoldia* 23(9):116.1963.

'Berlin' Origin unknown, possibly Gray Summit, MO.

**'Blauer Heinz'** H. Preissel in D*eutsche Baumschule* 516.1987.

'Bowles Blue' Royal Horticultural Society, Wisley, England.1986.

**'Broman'** Sheridan Nurseries, Toronto, Canada. Selected in 1936.

'Bruns' Inventory, Kingsville Nursery, Kingsville, MD.1968. = **'Heinrich Bruns'**

**'Bullata'** G. Kirchner in Petzold and Kirchner, *Arboretum Muscaviense* 194.1864.

**'Butterworth'** Catalog, Tingle Nurseries, Pittsville, MD 1958.

'Calfornia' Catalog, Moreau Landscape Nursery, East Colts Neck, N.J.1985.

**'Caucasica'** Hort. ex K. Koch, *Dendrologie* v.2,pt.2:476.1862.

**'Christiansen'** Catalog, Cary Brothers Nursery, Shrewsbury, MA 1957.

**'Clembrook'** E. Clements in *The Boxwood Bulletin* 8(2):20-22.1968.

**'Cliffside'** J. Baldwin in *The Boxwood Bulletin* 14(1):15.1974.

**'Columnaris'** Catalog, Visser's Nurseries, 132-9 Merrick Blvd., Springfield Gardens, Long Island N.Y.1960.

'Compacta' Catalog, Charles Dietriche, Angers, France.1892. ?

**'Conica'** Catalog, Siebenthaler Nurseries, Dayton, OH.136:10.1938.

**'Crispa'** Hort. ex K.Koch, *Dendrologie* v.2,pt.2:476.1872.

**'Croni'** Catalog, Monroe Nurseries, Monroe, MI.1955.

**'Cucullata'** Hort. ex K.Koch, *Dendrologie* v.2,pt.2:476.1872.

'Curly Locks' = *B. microphylla* Sieb. & Zucc. **'Curly Locks'**

**'Decussata'** L. Dippel, *Handbuch der Laubholzkunde* 3:82.1893.

**'Dee Runk'** C. Woltz in *The Boxwood Bulletin* 28(2):26.1988.

**'Denmark'** M. Gamble in *The Boxwood Bulletin* 28(2):28.1988.

**'Edgar Anderson'** M. Gamble in *The Boxwood Bulletin* 13(2):26-28.1973.

'Ed Wyckoff' H. Hohman, Kingsville Nursery, Kingsville, MD.1970.

'Elata' L. Dippel, *Handbuch der Laubholzkunde* 3:82.1893. =
  **'Angustifolia'**

**'Elegans'** L. Bailey, *Standard Cyclopedia of Horticulture* 601.1914.

**'Elegantissima'** Hort. ex K.Koch, *Dendrologie* v.2,pt,2:477.1872.

'Elegantissima Variegata' Catalog, Charles Dietriche, Angers, France. =
  **'Elegantissima'**

**'Fairview'** Catalog, Eastern Shore Nurseries, Inc., Easton, MD.49.1947.

**'Fastigiata'** F. Meyer in *Plant Explorations* ARS 34-9:91.1959.

'Fastigiata Hardwickensis' Catalog, Kingsville Nurseries, Kingsville,
  MD 1968

'Flavo-marginata' L. Dippel, *Handbuch der Laubholzkunde* 3:81.1893.
  = **'Marginata'**

'Flavo-variegatis' Beissner, Schelle and Zabel, *Handbuch der Laubholz-
  Benennung* 284.1903. = **'Aureo-variegata'**

'Flora Place' Edgar Anderson, St. Louis, MO. ?

'Fortunei Rotundifolia' Inventory, Royal Botanic Gardens, Kew,
  England.1961.

'Fruticosa' Duhamel, *Arbres & Arbustes*, Ed.2,i.t.24.1801-19. =
  **'Suffruticosa'**

'Fruticosa Foliis Variegata' F. Dietrch, *Vollstandiges Lexicon der
  Gartnerei und Botanik* 2:391.1802. = **'Suffruticosa Variegata'**

'Gigantea' V.Veillard in Duhamel, *Traité des Arbes et Arbrisseaux*
  ed.augm. 1:82.1835 = *B. balearica*

**'Glauca'** G. Kirchner in Petzold and Kirchner, *Arboretum Muscaviense*
  194.1864.

**'Glauca Marginata Aurea'** Catalog, F. Delaunay, Angers,
  France.1910.

**'Globosa'** Catalog, Siebenthaler Nurseries, Dayton, OH.136:10.1938.

'Golden' Commonly used in catalogs as a descriptive term. = **'Aureo-
  variegata'**

**'Graham Blandy'** L. Batdorf in *The Boxwood Bulletin* 25(1):8.1985.

**'Grandifolia'** J. Muller, Arg. in *De Candolle Prodromus* 16(1):19.1869.

**'Grand Rapids'** Catalog, Light's Tree Co., Richland, MI. 12:14.1948.

'Gray Summit' Inventory, Secrest Arboretum, Wooster, OH.

'Green Beauty' = *B. microphylla* var. *japonica* 'Green Beauty'

'Greenpeace' Braimbridge. Some Boxwoods in Cultivation. *The
  Plantsman. 15(4):248.1994*='Graham Blandy'

**'Handsworthiensis'** Fisher ex Henry in Elwes and Henry, *Trees of
  Britain and Ireland.* #7:1725.1913.

'Handsworthiensis Candelabra' Catalog, Kingsville Nurseries,
  Kingsville, MD.1967.

'**Handsworthii**' Hort. ex K. Koch, *Dendrologie* v,2.pt.2:476.1872.

'**Handsworthii Aurea**' Catalog, Visser's Nurseries, Springfield, Long Island, N.Y.1945

'**Hardwickensis**' Beissner, Schelle and Zabel, *Handbuch der Laubholz-Benennung* 283.1903.

'**Hardy Michigan**' Catalog, John Vermeulen and Son, Inc., Neshanic Station, N.J.1959.

'**Harmony Grove**' D. Wyman in *American Nurseryman* 107(7):57.1963.

'**Hendersonii**' Catalog, Lindley Nurseries, Greensboro, N.C.1958.

'**Heinrich Bruns**' F. Meyer, New Cultivars of Woody Ornamentals in *Baileya* 9(4):129.1961.

'**Henry Shaw**' M. Gamble in *The Boxwood Bulletin* 25(2):43-47.1985.

'**Hermann von Schrenk**' M. Gamble in *The Boxwood Bulletin* 14(2):31-ibc.1974.

'**Heterophylla**' V.Veillard in Duhamel, *Traité des Arbes et Arbrisseaux*, ed. augm.1:82.1835.

'Hillsboro' Rocknoll Nursery, Hillsboro, OH.1987.

'Hirsholmi' Kobenhavens Univ. Botanisk.1986 ?

'Holland' Catalog, Weller Nursery Co., Holland, MI.

'**Hood**' M. Gamble in *The Boxwood Bulletin* 26(3):64-67.1987.

'Horizontalis' Hillier Nurseries, *Manual of Trees and Shrubs* 1972. =
    '**Prostrata**'

'Humilis' K. Koch, *Syn.Fl.Germ.Helv.* ed.2,8:722.1844. =
    '**Suffruticosa**'

'**Inglis**' D. Wyman in *Arnoldia* 17(11):65.1957.

'Ipek' Edgar Anderson, St. Louis, MO.

'Japonica Aurea' Hillier Nurseries, *Manual of Trees and Shrubs* 1972. =
    '**Latifolia Maculata**'

'Jensen' Catalog, Moreau Landscape Nursery, Colts Neck, N.J.

'**Joe Gable**' Catalog, Kingsville Nursery, Kingsville, MD.1946

'**Joy**' M. Gamble in *The Boxwood Bulletin* 24(1):12-13.1984.

'Krossa-Livonia' Catalog, Kingsville Nursery, Kingsville, MD.1971.

'Lace' Smallscape Nursery, Suffolk, England.1992.

'Langley Pendula' Langley Boxwood Nursery, Hampshire, England.1991.

'Latifolia' Anonymous in *Annuals of Horticulture* 2:541.1847. =
    '**Bullata**'

'Latifolia Bullata' *Kew Handlist of Trees and Shrubs* 609.1902. =
    '**Bullata**'

'**Latifolia Macrophylla**' *Kew Handlist of Trees and Shrubs* 609.1902.

'**Latifolia Maculata**' *Kew Handlist of Trees and Shrubs* 131.1896.

'**Latifolia Marginata**' *Kew Handlist of Trees and Shrubs* 269.1925.
'Latifolia Pendula' Hillier Nurseries, Hants, England.1984.
'**Latifolia Nova**' *Kew Handlist of Trees and Shrubs* 609.1902.
'Lawson's Golden' Ashwood Nurseries, W. Midlands, England.1992.
'Ledifolia' Beissner, Schelle and Zabel, *Handbuch der Laubholz-Benennung* 283.1903. = '**Salicifolia**'
'Leptophylla' V.Veillard in Duhamel, *Traité des Arbes et Arbrisseaux* ed.augm. 1:82,t.23.Fig.3.1800 = '**Myrtifolia**'
'Liberty' Catalog, Schrader Peony Gardens, Liberty, IN.1955.
'Longifolia' G. Kirchner in Petzold and Kirchner, *Arboretum Muscaviense* 194.1864. = '**Angustifolia**'
'**Lynnhaven**' Greenbrier Farms, Inc., Norfolk, VA.1922.
'Macrocarpa' Beissner, Schelle and Zabel, *Handbuch der Laubholz-Benennung* 283.1903. = '**Macrophylla**'
'**Macrophylla**' Beissner, Schelle and Zabel, *Handbuch der Laubholz-Benennung* 283.1903.
'Macrophylla Glauca' Beissner, Schelle and Zabel, *Handbuch der Laubholz-Benennung* 283.1903. = '**Glauca**'
'Macrophylla Rotundifolia' Beissner, Schelle and Zabel, *Handbuch der Laubholz-Benennung* 283.1903. = '**Rotundifolia**'
'Maculatis' Beissner, Schelle and Zabel, *Handbuch der Laubholz-Benennung* 284.1903. = '**Marginata**'
'Maplewood' Bobbink Nurseries P.O. Box 124, Freehold, N.J.1987.
'**Marginata**' J.Loudon, *Arboretum et Fruticum Britannicum* III:1333.1838.
'**Mary Gamble**' J. Penhale in *The Boxwood Bulletin* 26(2):34-35.1986.
'**Memorial**' J. Baldwin in T*he Boxwood Bulletin* 6(4):ibc.1967.
'**Minima**' Beissner, Schelle and Zabel, *Handbuch der Laubholz-Benennung* 283.1903.
'**Minima Glauca**' Catalog, Charles Dietriche, Angers, France.1892.
'**Minor-aureo**' R. Weston, *Botanicus Universalis* 1:31.1770.
'Mirtifolia' Kingsville Nursery, Kingsville, MD.1968. = '**Myrtifolia**'
'**Mucronata**' Hortul. ex H. Baillon, *Monographie des Buxacées et des Stylocérées* 62.1859.
'**Myosotidifolia**' *Kew Handlist of Trees and Shrubs* 131.1896.
'**Myrtifolia**' *Catalog of Trees and Shrubs*, Gordon, Dermer and Edmonds Pl.6.1782.
'Myrtifolia Glauca' Beissner, Schelle and Zabel, *Handbuch der Laubholz-Benennung* 284.1903. = '**Suffruticosa Glauca**'
'Nana' V.Veillard in Duhamel, *Traité des Arbres et Arbrisseaux* ed. augm. 1:83.1835. = '**Suffruticosa**'
'**Natchez**' M. Gamble in *The Boxwood Bulletin* 26(3):62-63.1987.

'Navicularis' Catalog, Charles Dietriche, Angers, France.1892. =
**'Handsworthiensis'**

**'Newport Blue'** Catalog, Boulevard Nurseries, Newport, R.I.1941.

**'Nigricans'** P. Corbelli, *Dizionario di Floricultura* 232.1873.

**'Nish'** M. Gamble in *The Boxwood Bulletin* 14(4):61.1975.

'Northern Beauty' M. Dirr, *Manual of Woody Landscape Plants* 131.1983.

**'Northern Find'** D. Wyman in *Arnoldia* 23(5):87-88.1963.

**'Northern New York'** Inventory Beal-Garfield Botanic Garden, East Lansing, MI.1960.

**'Northland'** C.W.Stuart and Co., Newark, N.Y.1949.

**'Notata'** R. Weston, *Botanicus Universalis* 1:31.1770.

**'Oleaefolia'** L.H.Bailey, *Standard Cyclopedia of Horticulture* 601.1914.

'Oleaefolia Elegans' L.H.Bailey, *Standard Cyclopedia of Horticulture* 601.1914. = **'Pyramidalis'**

'Ohio' Bobbink Nurseries, Freehold, N.J.1987.

**'Pendula'** Catalog, Simon Louis 21.1869.

'Pendula Esveld' Langley Boxwood Nursery, Hampshire, England. 1991.

**'Ponteyi'** L. Dippel, *Handbuch der Laubholzkunde* 3:81.1893.

'Pride of Rochester' Girard Nurseries, Geneva, OH.1987.

'Prizren' Edgar Anderson, St. Louis, MO.1936.

**'Prostrata'** W. Bean, *Trees and Shrubs Hardy in the British Isles* 1:278.1914.

**'Pullman'** W. Pullman in *The Boxwood Bulletin* 11(2):20-21.1971.

**'Pyramidalis'** Catalog, Simon Louis 21.1869.

**'Pyramidalis Hardwickensis'** *Kew Handlist of Trees and Shrubs* 269.1925.

**'Pyramidalis Variegatis'** Catalog, Baudriller Nurseries, Angers, France.1880.

'Pyramidata' Inventory, Sanford Arboretum, TN.1932. = **'Pyramidalis'**

'Ransom' Appalachian Nurseries, Waynesboro, PA. 1964.

'Rochester' Girard Nurseries, Geneva, OH.1988.

**'Rosmarinifolia'** Hortul. ex H. Baillon, *Monographie des Buxacées et des Stylocérées* 61.1859.

'Rosmarinifolia Crispa' Beissner, Schelle and Zabel, *Handbuch der Laubholz-Benennung* 284.1903. = **'Suffruticosa Crispa'**

'Rosmarinifolia Fruticosa' P.Corbelli, *Dizionario di Floricultura* 1:232.1873. = **'Suffruticosa'**

'Rosmarinifolia Major' H. Baillon, *Monographie des Buxacées et des Stylocérées* 62.1859. = **'Rosmarinifolia'**

'Rosmarinifolia Minor' H. Baillon, *Monographie des Buxacées et des Stylocérées* 62.1859. = **'Suffruticosa'**

'**Rotundifolia**' H. Baillon, *Monographie des Buxacées et des Stylocérées* 61.1859.

'**Rotundifolia Aurea**' L. Dippel, *Handbuch der Laubholzkunde* 3:82.1893.

'**Rotundifolia Aureo-variegata**' Beissner, Schelle and Zabel, *Handbuch der Laubholz-Benennung* 284.1903.

'**Rotundifolia Maculata**' F. Meyer in *Plant Explorations* ARS 34-9:113b.1959.

'**Rotundifolia Minor**' Beissner, Schelle and Zabel, *Handbuch der Laubholz-Benennung* 284.1903.

'**Salicifolia**' Hort. ex K. Koch, *Dendrologie* v.2,pt.2:476.1872.

'**Salicifolia Elata**' Catalog, F.Delauney, Angers, France.1896.

'**Semi-elata**' Catalog, Charles Dietriche, Angers, France.1892.

'**Semperaurea**' B. Wagenknecht in *The Boxwood Bulletin* 7(1):1.1967.

'**Serbian Blue**' M. Gamble in *The Boxwood Bulletin* 14(4):61.1975.

'Silver Beauty' Burncoose & S. Down Nurseries, Cornwall, England. 1991.

'**Ste. Genevieve**' M. Gamble in *The Boxwood Bulletin* 11(1):1,15-16.1971.

'**Subglobosa**' Beissner, Schelle and Zabel, *Handbuch der Laubholz-Benennung* 283.1903.

'**Suffruticosa**' Linnaeus, *Species Plantarum* 983.1753.

'**Suffruticosa Alba Marginata**' Catalog, Brimfield Nurseries, Wethersfield, CT.1955.

'**Suffruticosa Aurea**' H. Baillon, *Monographie des Buxacées et des Stylocérées* 61.1859.

'**Suffruticosa Aureo-marginata**' Beissner, Schelle and Zabel, *Handbuch der Laubholz-Benennung* 284.1903.

'**Suffruticosa Crispa**' Beissner, Schelle and Zabel, *Handbuch der Laubholz-Benennung* 284.1903.

'**Suffruticosa Glauca**' Beissner, Schelle and Zabel, *Handbuch der Laubholz-Benennung* 284.1903.

'**Suffruticosa Maculata**' Beissner, Schelle and Zabel, *Handbuch der Laubholz-Benennung* 284.1903.

'Suffruticosa Myrtifolia' Beissner, Schelle and Zabel, *Handbuch der Laubholz-Benennung* 284.1903. = **'Myrtifolia'**

'Suffruticosa Nana' Catalog, W.T.Smith Co., Geneva, N.Y.1936. = **'Suffruticosa'**

'Suffruticosa Navicularis' Beissner, Schelle and Zabel, *Handbuch der Laubholz-Benennung* 284.1903. = **'Handsworthiensis'**

*A - Species, Cultivars and Hybrids*

'Suffruticosa Rosmarinifolia' Beissner, Schelle and Zabel, *Handbuch der Laubholz-Benennung* 284.1903. = **'Rosmarinifolia'**

'Suffruticosa Thymifolia' Catalog, Little and Ballantyne, Carlisle, England.1928. = **'Thymifolia'**

**'Suffruticosa Variegata'** R. Weston, *Botanicus Universalis* 1:31.1770.

'Suffruticosa Variegata Maculata' H. Baillon, *Monographie des Buxacées et des Stylocérées* 61.1859. = **'Argenteo-variegata'**

**'Tenuifolia'** Hortul. ex H. Baillon, *Monographie des Buxacées et des Stylocérées* 61.1859.

**'Thymifolia'** Beissner, Schelle and Zabel, *Handbuch der Laubholz-Benennung* 284.1903.

**'Thymifolia Variegata'** *Journal of the Royal Horticultural Society* 18:82.1895.

**'Undulifolia'** *Kew Handlist of Trees and Shrubs* 270.1902.

**'Vardar Valley'** D. Wyman in *Arnoldia* 17(7):42-44.1957.

'Variegata' Hort. ex Steud.Nom. ed II.i.242.

**'Varifolia'** Catalog, Kingsville Nursery, Kingsville, MD.1949.

**'Welleri'** Catalog, Weller Nursery Co., Holland, MI.1945.

'West Ridgeway' B. Blackburn, Willow Farm, Gladstone, N.J.1972.

'William Borek' Dept. of Horticulture, University of Vermont.1966.

'Woodland' Royal Botanic Gardens, Hamilton, Canada.1962.

'Yorktown' Selected by J. Baldwin, College of William & Mary, Williamsburg, VA.

---

***Buxus sinica*** (Rehd. & Wils.)Cheng **var.** *insularis* (Nakai)M.Cheng in *Acta Phytotax. Sin.*, 17(3):100.1979 & *Flora Reipublicae Popularis Sinicae* 45(1):37.1980.

'Arnold Promise' Miami Nursery, Tipp City, OH.1985.

'Cushion' B. Wagenknecht in *The Boxwood Bulletin* 7(1):1.1967 = **'Pincushion'**

'Dansville' University of Wisconsin-Madison.1986.

'Filigree' Langley Boxwood Nursery, Hampshire, England.1991.

**'Justin Brouwers'** P. Larson in *The Boxwood Bulletin* 29(1):3.1989.

**'Pincushion'** L. Batdorf in *The Boxwood Bulletin* 33(1):11.1993.

'Sunnyside' G. Krussmann, *Handbuch der Laubgeholze* 268.1976 = *B. microphylla* 'Sunnyside'

**'Staygreen'** Catalog, John Vermeulen and Son, Neshanic Station, N.J.1961.

**'Tall Boy'** B. Wagenknecht in *The Boxwood Bulletin* 7(1):1.1967.

**'Tide Hill'** D. Wyman in *Arnoldia* 17(11):64.1957.

**'Winter Beauty'** B. Wagenknecht in *The Boxwood Bulletin* 7(1):1.1967.

'**Wintergreen**' D. Wyman in *Arnoldia* 23(5):88.1963.

---

***Buxus wallichiana*** H. Baillon in *Monographie des Buxacées et des Stylocérées* 63-64.1859.

---

*Buxus* '**Green Gem**' B. Wagenknecht in *The Boxwood Bulletin* 7(1):1.1967.

*Buxus* 'Green Mound' Sheridan Nurseries, Georgetown, Ontario.1966.

*Buxus* '**Green Mountain**' B. Wagenknecht in *The Boxwood Bulletin* (7)1:1.1967.

*Buxus* '**Green Velvet**' B. Wagenknecht in *The Boxwood Bulletin* 7(1):1.1967.

# Appendix B
# Glossary

**Abiotic:** The nonliving factors of the environment that directly affect plant life.

**Adventitious roots:** Roots growing in unexpected positions, such as on a stem.

**Allelopathy:** When one plant directly or indirectly harms another, particularily when young, through the production of chemical compounds released into the environment.

**Apical meristems:** The growing point located at the tip of the stem. Flower, leaf, axillary bud primordia and the conductive tissues of the stem are initiated in the region.

**Axillary bud:** Laterally growing buds, those in the leaf axil, located immediately between the leaf petiole and the trunk.

**Chlorophyll:** The green plant pigment involved in the light reactions of photosynthesis.

**Clone:** A population of genetically identical plants.

**Conidium:** An asexual spore produced by many fungi, especially the ascomycetes.

**Cultivar:** Any variety or strain produced by horticultural methods that is not normally found in nature.

**Desiccation:** The loss of all water from a leaf or a portion of a plant.

**Drip line:** An imaginary line formed on the ground by the outermost branches. At this point, the rain would "drip" off the plant. It is also a measurement of the full width of the plant which is used to approximate root ball size.

**Electrical conductivity:** Used to measure the salinity hazard of irrigated water. If the sodium is more than 60% of the total calcium, magnesium, and sodium in the irrigated water, the diffusion of air and water into the soil is reduced.

**Feeder roots:** Small, thin-walled extensions of an epidermal cell of a root which increase the absorption of water and nutrients by roots.

**Gall (root):** A swelling, or rapid growth, of tissues resulting from the attack of certain parasites.

**Horticultural oil spray:** A highly refined oil, that is most effective against mites, scale insects, psyllids, as well as other small, soft-bodied, sucking pests. Dormant season sprays control overwintering eggs and inactive forms. Summer sprays (at half the dormant rate) control actively-feeding forms and eggs. Summer oil sprays are most effective for mite control. The foliage should be sprayed to run off, particularly the underside of leaves. There are a few precautions: do not spray if freezing temperatures are expected in 24 hours, when new leaves are opening in spring, when the leaves are wet, or the humidity is above 90%, or three weeks before or three weeks after a sulfur-containing spray.

**Hybrid:** A plant produced as the result of a cross between different species.

**Insecticidal soap:** Composed of 50% potassium salts of fatty acids (soap), it is safe for the applicator and is excellent in Integrated Pest Management programs where predators and parasites are cultivated. It has a short activity period and must be sprayed directly on most pests for good control.

**Larva(e):** The immature, wingless, often worm-like form of insects that hatch from an egg. They then progress to the pupa, then finally adult stage. It is the larval stage (feeding stage) that causes the greatest damage to plants.

**Mutation:** An offspring that exhibits variation from its parent's characteristics, caused by a heritable alteration in a gene.

**Nymph:** The preadult stage of insects with incomplete metamorphosis.

**Obligate parasite:** Organisms that require a host in order to grow.

**Oblong:** A leaf longer than broad, and with the sides nearly parallel most of their length.

**Obovate:** A leaf that is broader above the middle rather than below the middle.

**Ovipositor:** A specialized organ found on boxwood leaf miners. It extends from the end of the abdomen, forming a boring apparatus which creates holes in the leaves, in which eggs are placed.

**Open-pollinated:** A condition where either self- or cross-pollination has occurred. The pollen fertilizing the ovule may be from the same, or a different, plant.

**Parasite:** A plant (or animal) living in, on, or with another organism (the host) at whose expense the parasite is maintained, usually without immediately destroying the host.

**Parterre:** An ornamental and diversified design of flowering or foliage plants grown in beds.

**Pathogen:** A living organism that is able to cause disease.

**Permanent wilt:** The point at which the available water in the soil is so low that plant parts begin to wilt and will not recover.

**Petiole:** A short stalk at the base of the leaf. It attaches the leaf to the stem.

**pH:** A measure of the hydrogen ion concentration, [$H^+$], and thus the acidity or alkalinity of $H^+$ dissolved in solution. A pH value of 7 is neutral, less than 7 is acidic and a value above 7 is alkaline. One integer on the pH scale indicates a tenfold difference in $H^+$ concentration. For example, a pH of 5.5 is 100 times more acidic than a pH of 7.5.

**Pupation:** An intermediate, dormant stage of insects after the larva and before the adult.

**Pycnidium:** A structure in which the asexual spores are formed.

**Root prune:** The roots are pruned one or two years before transplanting to promote fibrous rooting within the future root ball. This is accomplished by inserting a flat-tip spade into the soil around the plant, slightly inside of where the root ball will be dug.

**Sepals:** One of the separate units of a calyx (the outer whorl of floral envelopes).

**Signs:** The disease or its parts seen on a host plant.

**Sport:** A plant or a part of a plant that shows deviation from the normal or parent type.

**Stamens:** The pollen-bearing organ of a flower having an anther and filament.

**Stippling:** Mite damage to the plant, resulting in discolored leaves that have tiny, light colored dots or short strokes.

**Stratification:** A pregermination treatment, exposing a seed to moisture in cold conditions for a specific time.

**Style:** The elongated part of the pistil between the ovary and stigma (the apical part of the pistil which receives the pollen grains).

**Symptoms:** The external and internal reactions or changes of a plant resulting from a disease.

**Temperate:** The middle latitude zone, lying between $23^1/2°$ and $66^1/2°$ north and south of the equator.

**Transpiration:** The release of water vapor from the plant occurring at the leaf surface.

**Variegation:** Green leaves that are partly white or yellow.

**Vascular cambium:** The tube-like tissues that carry food (products of photosynthesis) and water through the plant.

# Appendix C
# Bibliography

Agrios, George N. 1988. *Plant Pathology*. 3rd ed. Academic Press Inc., San Diego, California. 803 pp.

Batdorf, Lynn R. 1989. *Checklist of* Buxus *L.*, *The Boxwood Bulletin* 28(3):43-49.

Bell, D.K. 1967. *Etiology and Epiphytology of Root-Rot, Stem Necrosis, and Foliage Blight of Boxwood Caused by* Phytophthora parasitica *Dastur*. North Carolina Agricultural Experimental Station. Tech. Bulletin No. 177.

Brady, Nyle C. 1990. *The Nature and Properties of Soils*. 10th ed. Macmillan Publishing Company, New York, N.Y. 621 pp.

Brooklyn Botanic Garden Record. 1986. *American Gardens*. #111. Brooklyn Botanic Garden, Inc., Brooklyn, N.Y. 104 pp.

Hale, Maynard G. & David M. Orcutt. 1987. *The Physiology of Plants Under Stress*. John Wiley & Sons, New York. 206 pp.

Harris, Richard W. 1992. *Arboriculture*. 2nd ed. Prentice-Hall, Inc. Englewood Cliffs, New Jersey. 674 pp.

Hartl, Daniel L. 1994. *Genetics*. 3rd ed. Jones and Bartlett Publishers, Boston. 584 pp.

Hartmann, Hudson T. 1968. *Plant Propagation*. Prentice-Hall, Inc. Englewood Cliffs, New Jersey. 702 pp.

Jacob, Irene & W. Jacob. 1985. *Gardens of North America and Hawaii*. Timber Press. Portland, Oregon. 368 pp.

Johnson, Warren T. & Howard H. Lyon. 1991. *Insects That Feed on Trees and Shrubs*. Cornell University Press. Ithaca, N.Y. 560 pp.

Johnson, Warren T. & Howard H. Lyon. 1989. *Diseases of Trees and Shrubs*. Cornell University Press. Ithaca, N.Y. 575 pp.

Kays, Jonathan & Ethel Dutky. 1992. *Reducing Vole Damage to Plants in Landscapes, Orchards, and Nurseries*. Cooperative Extension Service, University of Maryland System. 5 pp.

Record, Samuel J. & George A. Garratt. 1925. *Boxwoods*. Yale University School of Forestry Bulletin #14. 81 pp.

Stone, Doris M. 1982. *The Great Public Gardens of the Eastern United States*. Pantheon Books. New York. 300 pp.

# Appendix D
# The American Boxwood Society

In 1961, the American Boxwood Society (ABS) was established by a group of professional and amateur boxwood enthusiasts. Headquarters are at the University of Virginia's Orland E. White Arboretum, Blandy Farm, Boyce, Virginia, the State Arboretum of Virginia. Through the years, the title of the ABS has been a bit misleading to some. The ABS is interested in all boxwoods, and is not limited to the so-called "American" boxwood.

The ABS is a non-profit organization devoted to educational objectives through its quarterly publication of *The Boxwood Bulletin*, and through popularizing the use of boxwood. The ABS also holds an annual meeting in May including a tour of its Boxwood Garden, lectures by authorities, and a sale of named boxwood plants.

The ABS is the International Registration Authority for Cultivated *Buxus* L. Mr. Lynn R. Batdorf serves as the registrar. The ABS sponsors and partially funds research projects on *Buxus* at several state universities. The Society established and supports the ABS Memorial Garden at Blandy Farm. This garden, landscaped with some 85 *Buxus* species and cultivars, was developed as a memorial planting honoring early leaders of the Society.

Individual memberships are available and include a subscription to *The Boxwood Bulletin*. Publications available are: *The Boxwood Handbook*; the *Boxwood Buyer's Guide* 4th edition, listing more than 100 nurseries where boxwood can be purchased; *International Registration List of Cultivated* Buxus L.; and indexes to *The Boxwood Bulletin*. Currently in progress is the *Manual of Boxwood Plants*, a publication which definitively describes and illustrates more than 350 of the temperate species and cultivars of boxwood.

For further information write: The American Boxwood Society, P.O. Box 85, Boyce, VA 22620-0085.

# Appendix E
# USDA Hardiness Map

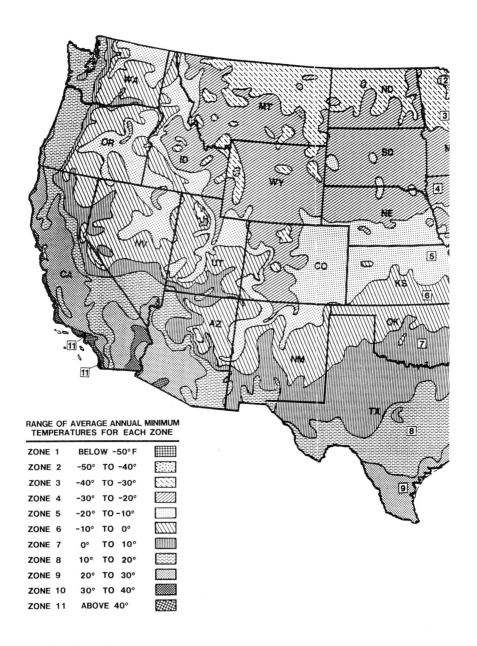

RANGE OF AVERAGE ANNUAL MINIMUM
TEMPERATURES FOR EACH ZONE

| | | |
|---|---|---|
| ZONE 1 | BELOW −50°F | |
| ZONE 2 | −50° TO −40° | |
| ZONE 3 | −40° TO −30° | |
| ZONE 4 | −30° TO −20° | |
| ZONE 5 | −20° TO −10° | |
| ZONE 6 | −10° TO 0° | |
| ZONE 7 | 0° TO 10° | |
| ZONE 8 | 10° TO 20° | |
| ZONE 9 | 20° TO 30° | |
| ZONE 10 | 30° TO 40° | |
| ZONE 11 | ABOVE 40° | |

*E - Hardiness Map*

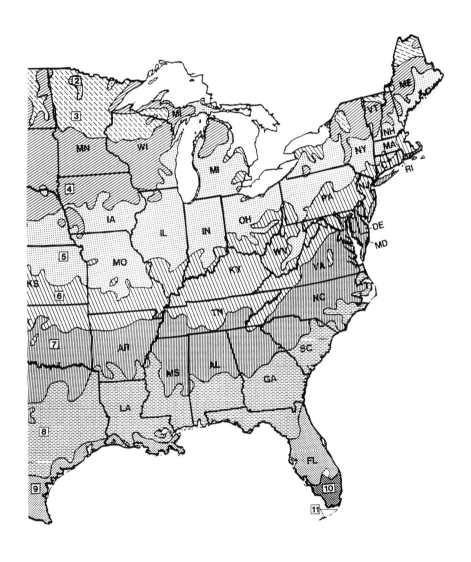

*E - Hardiness Map*